DIFFERENTIAL EQUATIONS

DIFFERENTIAL EQUATIONS

A Concise Course

H. S. BEAR
University of Hawaii

DOVER PUBLICATIONS, INC.
Mineola, New York

Bibliographical Note

This Dover edition, first published in 1999, is an unabridged and slightly corrected republication of the work originally published in the "Addison-Wesley Series in Mathematics" by Addison-Wesley Publishing Company, Inc., Reading, Mass., in 1962.

Library of Congress Cataloging-in-Publication Data

Bear, H. S. (Herbert Stanley)
    Differential equations : a concise course / H. S. Bear.
      p.   cm.
    Originally published: Reading, Mass. : Addison-Wesley Pub. Co., 1962, in series: Addison-Wesley series in mathematics.
    Includes index.
    ISBN 0-486-40678-4 (pbk.)
    1. Differential equations. I. Title.
QA372.B384   1999
515'.35—dc21
                                        99-13371
                                            CIP

Manufactured in the United States of America
Dover Publications, Inc., 31 East 2nd Street, Mineola, N.Y. 11501

PREFACE

This book is intended to serve as a text for a standard one-semester or two-term course in differential equations following the calculus. The author has given more than the usual emphasis to the mathematical explanations, in the conviction that there is little value in learning techniques by rote. There is more material presented here than can be covered in two terms, and the material ranges from routine calculations to moderately sophisticated theorems. This variety allows the instructor a fair degree of latitude in both the content and the level of the course to be taught.

The author would like to acknowledge his indebtedness to the University of Washington for the privilege of teaching from this text in a preliminary form, and to the many colleagues and students who have made helpful suggestions.

January, 1962   H. S. B.

CONTENTS

DIFFERENTIAL
EQUATIONS

CHAPTER 1

FIRST ORDER EQUATIONS

1-1 Introduction. A differential equation is an equation containing an "unknown" function and its derivatives. If the unknown is a function of one variable, the equation is called an *ordinary differential equation*, and if the unknown is a function of several variables so that the derivatives are partial derivatives, then the equation is a *partial differential equation*. We will be concerned exclusively with ordinary differential equations in this book. We will study equations such as

$$4x^3 f''(x) - 2x[f'(x)]^2 + f(x) - x^2 = 0, \tag{1}$$

and ask what *functions f* satisfy the equation *identically*.

As an example of how a differential equation arises naturally in a physical problem, consider an object of mass m falling through the air. Suppose that the object is only slightly more dense than air (e.g., a balloon) and that the air flows smoothly around the object as it falls. In this case the force due to air resistance is proportional to the speed. If we let $s(t)$ be the distance the object falls in time t, then the speed is $s'(t)$ and the acceleration is $s''(t)$. The forces acting on the object are the force of gravity, mg, and the force of air resistance, $ks'(t)$. According to Newton's law, force equals mass times acceleration, and the motion is described by the differential equation

$$mg - ks'(t) = ms''(t). \tag{2}$$

The highest order of the derivatives occurring in a differential equation is called the *order* of the equation. Thus a first order equation is any equation of the form

$$G(x, f(x), f'(x)) = 0.$$

We customarily use the symbol y for the unknown function, and for simplicity in writing we omit the independent variable from the expressions $y(x)$, $y'(x)$, etc. Hence a first order equation will appear as

$$G(x, y, y') = 0, \tag{3}$$

and the second order equations (1) and (2) would ordinarily be written

$$4x^3 y'' - 2x(y')^2 + y - x^2 = 0, \tag{4}$$

$$mg - ks' = ms''. \tag{5}$$

1

A function is a *solution* of a differential equation *on an interval I* if the function satisfies the equation identically on I. Thus a solution of (3) is any function y such that

$$G(x, y(x), y'(x)) \equiv 0,$$

where "$\equiv$" is used to indicate that the equality holds for all values of x in some interval. The function f defined by $f(x) = x^2$ is a solution of (1) on the whole line, since

$$4x^3(2) - 2x[2x]^2 + x^2 - x^2 \equiv 0.$$

Similarly, the function $s(t) = (mg/k)t$ is a solution of (2), since

$$mg - k\left(\frac{mg}{k}\right) \equiv m \cdot 0.$$

Notice that for any number c, the function $s(t) = (mg/k)t + c$ is also a solution of (2). This illustrates the fact that generally speaking a differential equation does not have just one solution, but an infinite family of solutions. This fact is familiar from the study of indefinite integration— i.e., the study of differential equations of the form

$$y' = f(x).$$

We will say that two equations are *equivalent* if any function which satisfies either equation also satisfies the other. That is, two equations are equivalent if they have the same sets of solutions. The following equations are easily seen to be equivalent:

$$\begin{align} y' + y &= x, \\ xy' &= x^2 - xy, \\ e^y(y' + y - x) &= 0. \end{align} \tag{6}$$

The equations

$$yy' + y^2 = xy, \tag{7}$$

$$(1 + y)(y' + y - x) = 0 \tag{8}$$

are not equivalent to the equations (6), since each has a solution which is not a solution of (6) (Problem 4).

PROBLEMS

1. Show that for any numbers c_1 and c_2 the function $s(t) = (mg/k)t + c_1 + c_2 e^{-(k/m)t}$ is a solution of (5).

2. For what values of c is $x^2 + c$ a solution of (4)?

3. Solve the following equations.

(a) $y'' = 0$ (b) $y' = \sec^2 x$ (c) $y' = \dfrac{1}{\sqrt{1 - x^2}}$

(d) $y'' = \dfrac{x + 1}{x}$ (e) $y' = \dfrac{1}{4 + x^2}$ (f) $y' = \dfrac{1}{x^2 - 9}$

4. Show that neither (7) nor (8) is equivalent to the equations (6). Are (7) and (8) equivalent?

5. Show that the functions given below are solutions of the corresponding differential equations. Find one more solution of each equation.

(a) $x^2 y' = xy - x$, $y = 2x + 1$ (b) $y'' + y' = 0$, $y = 3 - 2e^{-x}$

(c) $(x + 1)y' = -y$, $y = 1/(x + 1)$ (d) $y = (x + 1)y'$, $y = x + 1$

(e) $y'' + y = 0$, $y = \cos x$

ANSWERS

2. $c = 0$

3. (a) $y = c_1 x + c_2$ (b) $y = \tan x + c$

   (c) $y = \sin^{-1} x + c$ (d) $y = \tfrac{1}{2} x^2 + x \ln |x| + c_1 x + c_2$

   (e) $y = \tfrac{1}{2} \tan^{-1} \dfrac{x}{2} + c$ (f) $y = \tfrac{1}{6} \ln \left| \dfrac{x - 3}{x + 3} \right| + c$

4. (7) has the solution $y = 0$, (8) has the solution $y = -1$; no.

5. (a) $y = cx + 1$ (b) $y = c_1 + c_2 e^{-x}$

   (c) $y = c/(x + 1)$ (d) $y = c(x + 1)$

   (e) $y = c_1 \cos x + c_2 \sin x$

1–2 Variables separate. For any equation of the simple form

$$y' = f(x), \tag{1}$$

the solutions are obtained by integrating both sides. Thus

$$y = \int f(x) \, dx + c. \tag{2}$$

That is, to solve (1) we are required to find each function y whose derivative equals f. If F is any function such that $F'(x) = f(x)$, then any function y satisfying (1) differs from F by a constant. Therefore, all solutions of (1) must have the form

$$y(x) = F(x) + c. \tag{3}$$

Conversely, any function y of the form (3) obviously satisfies (1), and hence (3) characterizes the family of solutions of (1).

Aside from the familiar equations of the form (1), the simplest differential equations are those of the form

$$g(y)y' = f(x). \tag{4}$$

We will say that the *variables separate* in a differential equation if it is equivalent to an equation of the form (4). We agree that the equation

$$g(y)\, dy = f(x)\, dx \tag{5}$$

is equivalent to (4); i.e., we *define* y to be a solution of (5) if and only if y is a solution of (4).

As an example of an equation in which the variables separate, consider

$$3x^2 + 4y^3y' - 1 + y' = 0. \tag{6}$$

This is equivalent to

$$(4y^3 + 1)y' = 1 - 3x^2$$

and hence by definition to

$$(4y^3 + 1)\, dy = (1 - 3x^2)\, dx. \tag{7}$$

If we integrate both sides of (7), we obtain

$$y^4 + y = x - x^3 + c. \tag{8}$$

Two questions arise concerning this process. First, in what sense is (8) a solution of (6), since (8) is not a family of functions but a family of equations? The answer is that the solutions are the functions y which satisfy an equation of the form (8); i.e., the functions y which are defined implicitly by one of the equations (8). In many cases it is not possible to write explicit formulas for the solutions, and we regard the differential equation as solved if we can characterize the solutions in a manner not dependent on their derivatives. The second question is whether the family of equations (8) actually does characterize the solutions of (6); does every solution of (6) satisfy one of the equations (8), and is every function satisfying one of the equations (8) a solution of (6)? It is by no means obvious that the formal process of integrating both sides of (7) solves the differential equation; this process is justified by the following theorem.

THEOREM 1. *If f and g are continuous, and $F' = f$, $G' = g$, and y is any solution of $g(y)\, dy = f(x)\, dx$, then there is a constant c such that $G(y(x)) \equiv F(x) + c$. Conversely, any differentiable function y which satisfies $G(y) = F(x) + c$, for some c, is a solution of the equation $g(y)\, dy = f(x)\, dx$.*

Proof. The second assertion, that any function satisfying an equation $G(y) = F(x) + c$ is a solution of the equation $g(y)y' = f(x)$, is immediate from the chain rule for differentiation. Hence, we need only prove that there are no other solutions.

Let y be any solution of

$$g(y)\, dy = f(x)\, dx, \tag{9}$$

so that

$$g\big(y(x)\big)y'(x) = f(x) \tag{10}$$

for all x. Let $H(x) = G\big(y(x)\big)$, where G is any antiderivative of g. Then by the chain rule and the identity (10) we get

$$H'(x) = G'\big(y(x)\big)y'(x) = g\big(y(x)\big)y'(x) = f(x) = F'(x). \tag{11}$$

That is, H and F have the same derivative. It follows that there is a constant c such that $H(x) = F(x) + c$, and hence that $G\big(y(x)\big) = F(x) + c$.

EXAMPLE 1. (A) $2yy' = e^x$.

By the theorem, the solutions are just those functions y which satisfy $y^2 = e^x + c$, for some number c.

(B) $y' = x^2 y^2$.

Except for the possibility $y = 0$, this equation has the same solutions as $y^{-2}y' = x^2$. We check that $y = 0$ is a solution, and integrate to solve $y^{-2}y' = x^2$. The solutions of this equation are given by the equations $-y^{-1} = \frac{1}{3}x^3 + c$, or

$$y = \frac{-3}{x^3 + 3c}.$$

Since c is arbitrary, this family of functions can be written

$$y = \frac{-3}{x^3 + c}. \tag{12}$$

The solutions of $y' = x^2 y^2$ are therefore the function $y = 0$ and the functions (12).

(C) $x\, dx + y\, dy = 0$.

Integration gives $\frac{1}{2}x^2 + \frac{1}{2}y^2 = c$, which is the same family of equations as $x^2 + y^2 = c$. Note that no real-valued function y satisfies this last equation unless $c > 0$, so the solutions of (C) are also characterized by the equations $x^2 + y^2 = c^2$.

(D) $\dfrac{x+1}{y} dx = (x^2 + 1) \ln |y| dy.$

Multiplying both sides by y, and dividing by $x^2 + 1$, we obtain

$$\frac{x+1}{x^2+1} dx = y \ln |y| dy.$$

Integrating the left side above, we get

$$\int \left(\frac{x}{x^2+1} + \frac{1}{x^2+1} \right) dx = \tfrac{1}{2} \ln (x^2 + 1) + \tan^{-1} x.$$

The right side can be integrated by parts to give

$$\int y \ln |y| dy = \tfrac{1}{2} y^2 \ln |y| - \int \tfrac{1}{2} y^2 \frac{1}{y} dy$$

$$= \tfrac{1}{2} y^2 \ln |y| - \tfrac{1}{4} y^2.$$

The solutions of (D) are therefore the functions y defined implicitly by the equations

$$2 \ln (x^2 + 1) + 4 \tan^{-1} x = 2y^2 \ln |y| - y^2 + c.$$

A type of equation more general than that in which the variables separate is the *exact equation*

$$DF(x, y) = 0, \qquad \left(D = \frac{d}{dx} \right), \tag{13}$$

or

$$F_x(x, y) + F_y(x, y)y' = 0. \tag{14}$$

We agree that the following equation is equivalent to (13) or (14).

$$F_x(x, y) dx + F_y(x, y) dy = 0. \tag{15}$$

As an example of an exact equation we consider

$$2x - 2xyy' - y^2 = 0. \tag{16}$$

This equation can be written

$$D(x^2 - xy^2) = 0. \tag{17}$$

A function y is a solution of (16)—that is, by (17), a function such that

$$D(x^2 - x[y(x)]^2) \equiv 0,$$

if and only if there is a constant c such that

$$x^2 - x[y(x)]^2 \equiv c.$$

In conventional form, therefore, the solutions of (15) are given by the equations

$$x^2 - xy^2 = c.$$

An analysis similar to the foregoing, and to the proof of Theorem 1, shows that the solutions of (13) or (14) or (15) are those functions y which satisfy an equation of the form $F(x, y) = c$ (Problem 12).

EXAMPLE 2. (A) $xy' + 2yy' = x - y$.

This equation can be written

$$(xy' + y) + 2yy' - x = 0,$$

or

$$D[xy + y^2 - \tfrac{1}{2}x^2] = 0.$$

The solutions are therefore given by

$$xy + y^2 - \tfrac{1}{2}x^2 = c.$$

(B) $(y^2 + 3x^2)\, dx + 2xy\, dy = 0$.

This is equivalent to

$$(y^2 + 2xyy') + 3x^2 = 0,$$

or

$$D[xy^2 + x^3] = 0.$$

The solutions are determined by the equations

$$xy^2 + x^3 = c.$$

Exact equations are studied in more detail in Section 2–2 and Section 2–3.

PROBLEMS

Solve the following equations.

1. $y' = x/y$ 2. $xy\, dy + (x^2 + 1)\, dx = 0$
3. $(1 - y^2)\, dx - xy\, dy = 0$ 4. $xy + \sqrt{1 + x^2}\, y' = 0$
5. $(y + 1)\, dx + (y - 1)(1 + x^2)\, dy = 0$
6. $(2x + 1)y' + y = 0$
7. $y' - x^2 = x^2 y$ 8. $y\, dx + x\, dy = 0$
9. $x^2\, dy + 2xy\, dx = x^2\, dx$ 10. $(y^2 + 2x)\, dx + 2xy\, dy = 0$

11. Show that any equation of the form $f(x)\,dx + g(y)\,dy = 0$ is exact.

12. Show that the solutions of (14) are the functions y defined implicitly by an equation of the form $F(x, y) = c$.

13. Show that the equation $y' + y = 0$ becomes exact if multiplied by e^x, and solve it.

14. Show that the equation $y - xy' = 2y^3y'$ becomes exact if multiplied by $1/y^2$, and solve it.

15. Find each positive function whose derivative is the reciprocal of the function.

16. Find all functions such that the square of the function plus the square of the derivative is a given constant A^2.

17. Find all functions such that the derivative is the square of the function.

18. Find all functions with derivative one more than the square of the function.

ANSWERS

1. $y^2 - x^2 = c$
2. $x^2 + y^2 + 2\ln|x| = c$
3. $x^2(1 - y^2) = c$
4. $\sqrt{1 + x^2} + \ln|y| = c, \quad y = 0$
5. $\tan^{-1} x + y - 2\ln|y + 1| = c, \quad y = -1$
6. $(2x + 1)y^2 = c$
7. $3\ln|1 + y| = x^3 + c, \quad y = -1$
8. $xy = c$
9. $3x^2y = x^3 + c$
10. $xy^2 + x^2 = c$
13. $y = ce^{-x}$
14. $x/y = y^2 + c, \quad y = 0$
15. $y = \sqrt{2x + c}$
16. $y = A \sin (x + c)$
17. $y = -1/(x + c), \quad y = 0$
18. $y = \tan (x + c)$

1–3 Geometric interpretation of first order equations. In Section 1–1 we saw how a differential equation arises in the attempt to solve the physical problem of a falling body. Consider now a geometrical problem. What functions, with graphs in the first quadrant, have the property that for every tangent line the point of tangency bisects the segment cut off by the coordinate axes (Fig. 1–1)? For such a curve, the tangent at the point (x, y) must have slope $-2y/2x = -y/x$. Hence any such function must satisfy the differential equation

$$y' = \frac{-y}{x}, \qquad y > 0, \qquad x > 0, \tag{1}$$

and conversely, a function satisfying (1) satisfies the conditions of the problem. The solutions of (1) (Problem 8, Section 1–2) are the functions $y = c/x, \ c > 0$.

Any first order differential equation $y' = F(x, y)$ admits a geometric interpretation similar to that illustrated in the problem above. The equation $y' = F(x, y)$ prescribes a tangent line at each point of the plane:

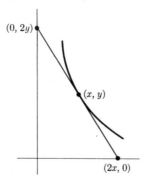

FIGURE 1-1

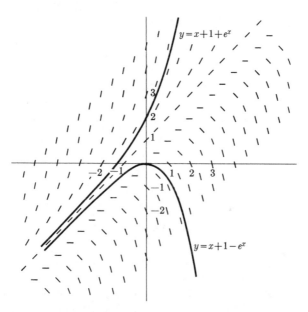

FIG. 1-2. Tangent field for $y' = y - x$.

the line through (x, y) whose slope is $F(x, y)$. The question posed by the differential equation $y' = F(x, y)$ is: What functions y have the tangents prescribed by the function F at each point of their graphs? We can visualize the situation by drawing short segments of the tangent lines given by the equation at various points of the plane. Figure 1-2 shows such a "tangent field" for the equation

$$y' = y - x, \tag{2}$$

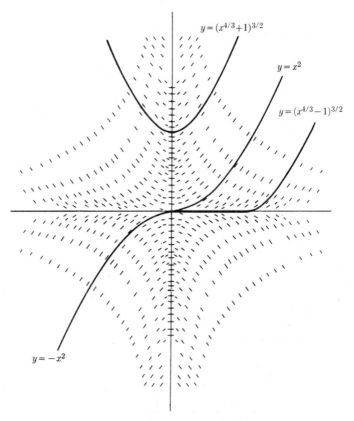

FIG. 1-3. Tangent field for $y' = 2x^{1/3}y^{1/3}$.

with graphs of the solutions $y = x + 1 + e^x$ and $y = x + 1 - e^x$. An easy way to construct a graph of the tangent field is to draw all the segments with a given slope at the same time. This amounts to graphing the curves $F(x, y) = c$ [the lines $y - x = c$ for (2)] and marking segments with slope c at intervals along the curves. Notice that the linear solution, $y = x + 1$, can be found by inspection of the tangent field for (2).

Figure 1-2 suggests that there is a solution, and only one solution, through any given point in the plane. The solutions of (2) are the functions

$$y = x + 1 + ce^x, \tag{3}$$

and it is easy to check that for any point (a, b), there is exactly one number c such that $b = a + 1 + ce^a$. In other words, there is exactly one solution curve through each point of the plane. It is true in general that if F is continuous, there is a solution to $y' = F(x, y)$ through each point (a, b). The solution through a given point is *not* necessarily unique without further assumptions on F.

Let us examine in some detail the equation

$$y' = 2\sqrt[3]{xy} = 2x^{1/3}y^{1/3}. \tag{4}$$

The tangent field for (4) with the tangent segments drawn along the hyperbolas $2\sqrt[3]{xy} = c$ is shown in Fig. 1–3. On any interval on which a solution is never zero, we can separate variables in (4) and obtain the equivalent form

$$y^{-1/3}y' = 2x^{1/3}. \tag{5}$$

According to Theorem 1, Section 1–2, we obtain an equivalent equation by integrating both sides to get

$$\tfrac{3}{2}y^{2/3} = \tfrac{3}{2}x^{4/3} - \tfrac{3}{2}c,$$

or

$$y = \pm(x^{4/3} - c)^{3/2}, \qquad (x^{4/3} > c). \tag{6}$$

Thus we have shown that any solution of (4) must have the form (6) on any interval on which it is nonzero.

The curves (6) which intersect the x-axis ($c \geq 0$) are tangent to the x-axis at the point of intersection, and hence satisfy (4) at this point (since $y = y' = 0$). The function identically zero is a solution of (4) on any interval. Hence any function whose graph follows one of the curves (6) to the x-axis, and continues along the x-axis until it leaves by another of the curves (6), is a solution of (4). There is a solution through every point, and there are infinitely many solutions through any point on the x-axis. As an example, the function

$$y(x) = \begin{cases} -x^2, & x \leq 0 \\ 0, & 0 < x < 1 \\ (x^{4/3} - 1)^{3/2}, & 1 \leq x \end{cases}$$

is a solution of the differential equation. The functions $y(x) = x^2$, $y(x) = -x^2$, $y(x) = 0$ are other solutions through the origin.

Problems

1. Find all curves such that for any point on the curve, the area bounded by the ordinate line to the point, the x-axis, and the tangent line at the point is $\tfrac{1}{2}$. Find the one of these curves through $(0, 1)$.

2. Construct a tangent field for $y' = 1 + x + y$. Find a linear solution by inspection of the tangent field and check it in the equation.

3. Consider the curves such that the tangent at each point (x, y) of the curve intersects the y-axis at $(0, -y)$. Draw the tangent field in the upper half-plane for the implied differential equation. [*Hint:* The tangent segment at each point on $y = 1$ must point toward $(0, -1)$, etc.] Write the differential equation, find all such curves, and graph several of them on the tangent field.

4. (a) Construct a tangent field for $y' = 2\sqrt{y}$. (Note that necessarily $y \geq 0$ and $y' \geq 0$.)

   (b) Find all solutions of $y' = 2\sqrt{y}$ as in the treatment of Eq. (4) in the text.

   (c) How many solutions are there through each point on the x-axis? Each point off the x-axis? Explain.

5. Describe the solutions of $y' = 3xy^{1/3}$.

ANSWERS

1. $y = 1/(c \pm x)$, $y = 1/(1 \pm x)$
2. $y = -x - 2$
3. $y = cx^2$
4. The solutions are $y = 0$, and the functions y_c, where $y_c(x) = 0$ if $x \leq c$ and $y_c(x) = (x - c)^2$ if $x \geq c$. There are infinitely many solutions through each point on the x-axis, and one solution through each point in the upper half plane.
5. The solutions are those continuous functions whose graphs lie always on one of the curves $y = (x^2 + c)^{3/2}$, $y = 0$, or $y = -(x^2 + c)^{3/2}$.

1–4 Existence and uniqueness theorem. The examples in the preceding section lead us to ask for a general statement about the existence and uniqueness of solutions of the first order equation

$$y' = F(x, y). \tag{1}$$

The question we ask is this: If the function F satisfies certain conditions near a point (a, b), is there a solution curve passing through (a, b)? If so, is there exactly one solution through (a, b)? In other words, we would like to know that there is exactly one function y such that $y(a) = b$, and $y'(x) \equiv F(x, y(x))$ on some interval around a.

There are several reasons for asking questions such as these. First of course is the fact that before we look for solutions of a differential equation, we would like to be assured that there are solutions to be found. The uniqueness part of the theorem is also of more than academic interest. We will illustrate below how the uniqueness of the solution through each point can be used to show that a given family of solutions contains *all* solutions.

We state here one convenient form of an existence and uniqueness theorem for Eq. (1). The hypothesis we use is the continuity of F and $F_y = \partial F/\partial y$ near a given point (a, b). In geometric terms, the fact that

F is continuous means that the tangent field determined by F changes direction smoothly as the point (x, y) moves in the plane near (a, b). We will see in Chapter 6 that the continuity of F_y rules out the sort of behavior illustrated in Fig. 1–3, where solution curves are tangent to each other, and there can be several solutions through a given point.

THEOREM 1. *If F and F_y are continuous on some square S centered at (a, b), then there is a function y defined on some interval I around a such that $y(a) = b$ and $y'(x) = F(x, y(x))$ for all x in I. Moreover, if g is any function such that $g(a) = b$ and $g'(x) = F(x, g(x))$ for x in I, then $g(x) = y(x)$ for x in I.*

This theorem is proved in Chapter 6.

Note that the existence theorem is local in character. If F is well behaved near (a, b), then there is, on *some* interval I around a, a solution whose graph goes through (a, b). The interval I may be small, even if F and F_y are continuous on the whole plane (see Problem 1).

EXAMPLE 1. Consider the equation (cf. Fig. 1–2)

$$y' = y - x. \tag{2}$$

This is of the form (1), with $F(x, y) = y - x$, and $F_y(x, y) = 1$. The functions F and F_y are continuous everywhere, so there is a unique solution through each point (a, b). It is easy to verify (Problem 2) that for any number c, the function

$$y = ce^x + x + 1 \tag{3}$$

is a solution of (2). For any given point (a, b) there is a number c such that

$$b = ce^a + a + 1.$$

That is, there is a solution from the family (3) through any given point. Since by Theorem 1 there is only one solution through any point, the solution through a given point must be the solution from the family (3). In other words, the family (3) contains all solutions of (2).

EXAMPLE 2. Consider the equation (cf. Fig. 1–3)

$$y' = 2x^{1/3}y^{1/3}. \tag{4}$$

Here $F(x, y) = 2x^{1/3}y^{1/3}$, and $F_y(x, y) = \frac{2}{3}x^{1/3}y^{-2/3}$. The function F is continuous everywhere, but F_y is continuous only at points (x, y) with $y \neq 0$. There is a solution of (4) through every point (a, b), but not necessarily a unique solution. If we restrict our attention to a square wholly above or below the x-axis, then there is *in the square* a unique solution curve through each point.

There are existence theorems analogous to Theorem 1 for equations of order higher than one. For example, for the second order equation

$$y'' = F(x, y, y') \qquad (5)$$

the following is true. If F is a function of three variables which is continuous near the point (a, b, c) and has continuous partial derivatives with respect to the second and third variables, then there is a unique function y defined on some interval around a which satisfies the differential equation

$$y''(x) \equiv F(x, y(x), y'(x)),$$

and satisfies the initial conditions

$$y(a) = b, \qquad y'(a) = c.$$

PROBLEMS

1. Solve the equation $y' = 100 + y^2$. Show that even though $F(x, y) = 100 + y^2$ and $F_y(x, y) = 2y$ are continuous everywhere, no function is a solution of the equation on any interval longer than $\pi/10$.

2. Verify that each function in the family (3) is a solution of the Eq. (2).

3. (a) Use Theorem 1 to show that $y' = 1 + x - y$ has exactly one solution through each point.

   (b) Verify that $y = x + ce^{-x}$ is a solution for every number c.

   (c) Show that for every point (a, b) there is a number c so that the solution $y = x + ce^{-x}$ satisfies $y(a) = b$.

   (d) Conclude from (a), (b), and (c) that the family $y = x + ce^{-x}$ is the family of all solutions of $y' = 1 + x - y$.

4. Proceed as in Problem 3 to show that $y' = 3x^3 - 2xy$ has a unique solution through each point, and that $y = \frac{3}{2}x^2 - \frac{3}{2} + ce^{-x^2}$ is the family of all solutions.

5. Solve the following equations. Write the equations in the form $y' = F(x, y)$ and show by Theorem 1 that there is a unique solution through the given point (a, b). Find the particular solution through the given point.

   (a) $x\,dy + y\,dx = x^2\,dx$; $(a, b) = (1, 0)$

   (b) $(1 + 2x^2)\,dy - x(1 + y^2)\,dx = 0$; $(a, b) = (0, 1)$

   (c) $(1/x)\,dy - 4\sqrt{1 + y}\,dx = 0$; $(a, b) = (2, 3)$

   (d) $dy = 2xy\,dx$; $(a, b) = (1, 0)$

6. Solve the following second order equations and find the solution which satisfies the given initial conditions.

   (a) $y'' = y'$; $y(0) = 2$, $y'(0) = 1$

   (b) $2y'y'' = 1$; $y(1) = 5$, $y'(1) = 2$

   (c) $xy'' + y' = x^2$; $y(1) = 1$, $y'(1) = 1$ [cf. Problem 5(a)]

ANSWERS

1. $y = 10 \tan (10x + c)$

5. (a) $3xy = x^3 + c$; $c = -1$
   (b) $4 \tan^{-1} y = \ln (1 + 2x^2) + c$; $c = \pi$
   (c) $\sqrt{1 + y} = x^2 + c$ and $y = -1$; $c = -2$
   (d) $y = ce^{x^2}$; $c = 0$

6. (a) $y = c_1 e^x + c_2$, $y = e^x + 1$
   (b) $y = \frac{2}{3}(x + c_1)^{3/2} + c_2$, $y = \frac{2}{3}(x + 3)^{3/2} - \frac{1}{3}$
   (c) $y = \frac{1}{9}x^3 + c_1 \ln |x| + c_2$, $y = \frac{1}{9}x^3 + \frac{2}{3} \ln |x| + \frac{8}{9}$

1-5 Families of curves and envelopes. We have seen that the set of solutions of a differential equation can frequently be expressed as a single formula with one or more arbitrary constants, or parameters. We will consider now the problem converse to that of solving a differential equation; namely, given a family of functions, is there a differential equation for which this family is the set of solutions? In other words, can we characterize a family of functions with a single differential equation?

If we have a family of functions

$$y = F(x, c), \tag{1}$$

then each function in the family (1) must also identically satisfy

$$y' = F_x(x, c). \tag{2}$$

If we can eliminate c from these equations—that is, derive from (1) and (2) an equation

$$G(x, y, y') = 0,$$

then this is a differential equation which is satisfied by every function in the family (1).

Consider the family of functions

$$y = 1 + cx. \tag{3}$$

Each of the functions (3) satisfies $y' = c$, and hence the equation

$$y = 1 + xy'. \tag{4}$$

Equation (4) is therefore a differential equation whose solutions contain all the functions (3). By separating variables and solving, one shows that (4) has no solutions other than those from the family (3).

A two-parameter family of functions will be represented by a second order equation. Differentiating the formula

$$y = c_1 x + c_2 e^x \tag{5}$$

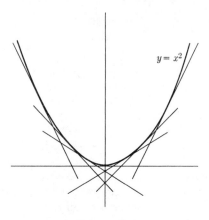

FIGURE 1–4

twice, we see that every function (5) satisfies the three equations

$$\begin{aligned}
y &= c_1 x + c_2 e^x, \\
y' &= c_1 + c_2 e^x, \\
y'' &= c_2 e^x.
\end{aligned} \tag{6}$$

We can solve the third equation for c_2, then the second for c_1, and substitute these values in the first equation to obtain

$$y = (y' - y'')x + y''. \tag{7}$$

We will be able to show in Chapter 3 that there are no other solutions of (7) and hence that (7) does characterize the family (5).

It is not always the case, as we show next, that the differential equation we derive from a family of functions has only solutions from this family. Consider the family of lines tangent to the curve $y = x^2$ (Fig. 1–4). The tangent line at (c, c^2) has the equation $y - c^2 = 2c(x - c)$, or

$$y = 2cx - c^2. \tag{8}$$

To obtain the differential equation of the family (8), we differentiate and get $y' = 2c$. The differential equation of (8) is, therefore,

$$y = xy' - \tfrac{1}{4}(y')^2. \tag{9}$$

The equation (9) prescribes a slope for a solution curve at each point of the plane. At each point (x, x^2), the curve $y = x^2$ has the same slope as its tangent line, which is a solution of (9) by virtue of the way (9) was

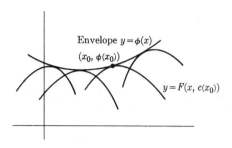

Envelope $y = \phi(x)$
$(x_0, \phi(x_0))$
$y = F(x, c(x_0))$

FIGURE 1–5

derived. Hence $y = x^2$ is a solution of (9), and (9) does *not* characterize the family of lines (8).

The situation encountered above occurs whenever a family of curves has an envelope. We will say that the curve $y = \phi(x)$ is an *envelope* of the family of curves $y = F(x, c)$ if at each point on $y = \phi(x)$ there is a curve from the family which is tangent there (Fig. 1–5). That is, for each point $(x_0, \phi(x_0))$ there is a value of c, $c = c(x_0)$, such that $y = \phi(x)$ and $y = F(x, c(x_0))$ are tangent at $(x_0, \phi(x_0))$. We also suppose that $c'(x)$ exists and $c'(x) \neq 0$. With these assumptions, we can give a simple test for determining whether a family $y = F(x, c)$ has an envelope, and we can write a formula for the envelope if there is one. The facts are stated in the following theorem.

THEOREM 1. *The family of curves $y = F(x, c)$ has the envelope $y = \phi(x)$ if and only if there is a function c such that $F_c(x, c(x)) \equiv 0$, and $\phi(x) = F(x, c(x))$.*

To restate the theorem: The envelope, if any, is the curve $y = F(x, c(x))$, where c is defined implicitly by the equation $F_c(x, c) = 0$. The equation of the envelope is obtained by eliminating c between the equations

$$y = F(x, c), \qquad 0 = F_c(x, c).$$

Proof. Assume that $y = \phi(x)$ is an envelope of the family $y = F(x, c)$. Let c be the function such that $y = \phi(x)$ and $y = F(x, c(x_0))$ are tangent at $(x_0, \phi(x_0))$, for each x_0. Since the curves $y = \phi(x)$ and $y = F(x, c(x_0))$ meet at $(x_0, \phi(x_0))$, we have $\phi(x_0) = F(x_0, c(x_0))$, for each x_0; that is, we have the identity

$$\phi(x) = F(x, c(x)). \tag{10}$$

The condition that the curves be tangent implies that for each x_0

$$\phi'(x_0) = D\,F(x, c(x_0))|_{x=x_0} = F_x(x, c(x_0))|_{x=x_0} = F_x(x_0, c(x_0)).$$

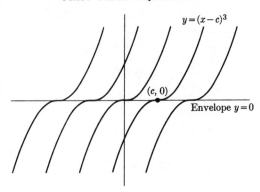

FIGURE 1-6

Hence we have the identity

$$\phi'(x) = F_x(x, c(x)).$$ (11)

If we compute ϕ' from (10), we get

$$\phi'(x) = F_x(x, c(x)) + F_c(x, c(x))c'(x).$$ (12)

Comparing (11) and (12) we have, since $c'(x) \neq 0$,

$$F_c(x, c(x)) = 0.$$ (13)

Formulas (13) and (10) are the assertions of the theorem, if we assume that there is an envelope.

The arguments above can be reversed to show that (13) and (10) are sufficient for $y = \phi(x)$ to be an envelope of the family $y = F(x, c)$. This part of the proof is left as a problem (Problem 8).

EXAMPLE 1. Consider again the family

$$y = 2cx - c^2.$$

Equation (13) for this example is

$$2x - 2c = 0,$$

which defines the function $c(x) = x$. The envelope is therefore [Formula (10)]

$$y = 2x \cdot x - x^2 = x^2.$$

EXAMPLE 2. Consider the family (Fig. 1-6)

$$y = (x - c)^3.$$

Partial differentiation with respect to c gives [Eq. (13)]

$$0 = 3(x - c)^2,$$

and hence $c = x$. There is, therefore, the envelope

$$y = (x - x)^3 = 0.$$

For a family of curves given in the form

$$G(x, y, c) = 0 \qquad (14)$$

it can be shown that the envelopes, if any, are obtained by eliminating c from (14) and

$$G_c(x, y, c) = 0. \qquad (15)$$

EXAMPLE 3. $(x - c)^2 + y^2 = 1.$

We eliminate c between the equations

$$(x - c)^2 + y^2 = 1, \qquad 2(x - c)(-1) = 0$$

to obtain the equation of the envelope curves, $y^2 = 1$.

Summary. To find the differential equation of an n-parameter family of curves, differentiate the given equation n times and eliminate all the constants.

To find the envelope of a one-parameter family of curves, eliminate the parameter between the given equation and that obtained from it by partial differentiation with respect to the parameter.

PROBLEMS

1. Solve Eq. (4) and show that there are no solutions other than the functions (3).

2. Find the differential equations of these families.

(a) $y = c + cx$ (b) $y = c_1 + c_2 e^x$
(c) $y = x^2 + cx + c^2$ (d) $y = c_1 + c_2 x + c_3 x^2$

3. Considering (6) as three simultaneous linear (algebraic) equations in c_1 and c_2, the condition that the equations be consistent is that the following determinant be zero:

$$\begin{vmatrix} x & e^x & y \\ 1 & e^x & y' \\ 0 & e^x & y'' \end{vmatrix} = 0.$$

Expand the determinant to obtain (7).

4. Proceed as in Problem 3 to find the differential equation of the family $y = c_1 e^{-x} + c_2 e^x - x$.

5. Show that each of the parabolas $y = x^2 + cx + \frac{1}{4}(1 - c)^2$ is tangent to $y = x$. Find a first order differential equation for the family and verify that $y = x$ is also a solution.

6. Find the envelope of the family $y = e^{x-c} + c$. Graph several of the curves and the envelope. Find the differential equation of the family and verify that the envelope is also a solution.

7. Find the envelope of the family $y = 3c^2 x - 2c^3$.

8. Show that if c is a differentiable function such that $c'(x) \neq 0$ and $F_c(x, c(x)) = 0$, then $y = F(x, c(x))$ is an envelope of the family $y = F(x, c)$. (See the proof of Theorem 1.)

9. Show that there is no envelope to the family of solutions of $g(y)y' = f(x)$. [*Hint:* Write the family of solutions and show that (15) is impossible.]

10. Solve $4y = y'^2$ by separating variables. Show that the family of solutions has an envelope which is a solution. Reconcile these facts with Problem 9.

11. Show that every solution of (9) everywhere agrees with one of the functions (8), or with $y = x^2$. [*Hint:* Differentiate (9) and examine the resulting second order equation. Show that either y' is constant, and y is in the family (8), or $y = x^2 + c$ with c necessarily zero.]

Answers

2. (a) $y = y' + xy'$ (b) $y'' = y'$
   (c) $y = 3x^2 - 3xy' + y'^2$ (d) $y''' = 0$

4. $y'' - y - x = 0$

5. $y = x^2 + (y' - 2x)x + \frac{1}{4}(1 - y' + 2x)^2$

6. $y = x + 1;\ y = y' + x - \ln y'$

7. $y = x^3$

10. $y = (x - c)^2,\ y = 0$; the separated form is not equivalent to the original.

1–6 Clairaut equations. Now we consider the differential equation of a particularly simple family of curves—a family of straight lines. The differential equation of such a family turns out to have an easily recognizable form. We have, therefore, a class of differential equations for which a family of solutions can be found by inspection.

The families of lines we consider are those which can be written in the form

$$y = cx + g(c). \tag{1}$$

Differentiating (1), we obtain $y' = c$, and hence each of the functions (1) must satisfy

$$y = xy' + g(y'). \tag{2}$$

Any equation of the form (2) is called a *Clairaut equation*. For any such equation, we can write down by inspection the one-parameter family (1) of solutions. There may be solutions of (2) not contained in the family (1), and we know that any envelope of (1) would be such a solution.

EXAMPLE 1. The equation

$$y = xy' - \tfrac{1}{4}(y')^2$$

is a Clairaut equation and therefore has the solutions

$$y = cx - \tfrac{1}{4}c^2.$$

This family of lines has the envelope (cf. Section 1-5) $y = x^2$, which is also a solution.

EXAMPLE 2. $y = xy' + 2 + y'$

This equation is of the form (2), so we can write by inspection the solutions

$$y = cx + 2 + c.$$

Differentiating this equation with respect to c we get $0 = x + 1$, which is obviously not satisfied by any function c. The family of solutions has no envelope.

EXAMPLE 3. The equation

$$(y - y'x)^2 = 2y' + 1$$

is essentially a Clairaut equation and has the solutions

$$(y - cx)^2 = 2c + 1.$$

To check for an envelope, we differentiate with respect to c and obtain

$$2(y - cx)(-x) = 2, \qquad c = \frac{1 + xy}{x^2}.$$

The family has the following envelope, which is a solution of the differential equation,

$$y = -\frac{1 + x^2}{2x}.$$

We have left unresolved the question of whether there are solutions of (2) other than the family (1) and its envelopes. We can obtain other solutions of (2) by piecing together functions from the family and the envelope at points of tangency. For example, the function

$$y = \begin{cases} 0, & x \le 0 \\ x^2, & 0 \le x \le 1 \\ 2x - 2, & 1 \le x \end{cases}$$

is a solution of the equation of Example 1. The question is, therefore, whether every solution of (2) must everywhere coincide either with one of the functions (1), or with an envelope of the family (1). We show in the following theorem that this is the case.

THEOREM 1. *If y is any solution of $y = xy' + g(y')$, then y is everywhere equal either to one of the functions $y = cx + g(c)$, or to an envelope of this family.*

Proof. If y satisfies the equation (2), then by differentiating and simplifying, we see that y also satisfies the identities

$$\begin{aligned} y' &= xy'' + y' + g'(y')y'', \\ y''[x + g'(y')] &= 0. \end{aligned} \tag{3}$$

On any interval on which $y'' = 0$, y' is some constant c, and from (2) we conclude that

$$y = cx + g(c). \tag{4}$$

If the other factor in (3) is zero,

$$x + g'(y') = 0, \tag{5}$$

then there is a function $c = c(x)$ satisfying

$$0 = \frac{\partial}{\partial c}\,[cx + g(c)] = x + g'(c); \tag{6}$$

in fact, the function $c(x) = y'(x)$ satisfies (6), by (5). Equation (6) is just the condition of Theorem 1, Section 1–5, that the family (4) have an envelope. By the theorem, the envelope function ϕ is given by

$$\phi(x) = c(x)x + g(c(x)) = xy'(x) + g(y'(x)). \tag{7}$$

Comparing (7) and the differential equation,

$$y = xy' + g(y'),$$

we see that the solution y is an envelope of the family (4), if y' satisfies (5).

To summarize, a function y which satisfies (2) is one of the functions $y = cx + g(c)$ on any interval where $y'' = 0$, and is an envelope of this family where $x + g'(y') = 0$.

The technique of the proof above extends to equations other than the Clairaut equation. Suppose we have any first order equation which can be solved for y—that is, written in the form

$$y = F(x, y'). \tag{8}$$

The solutions of (8) are given explicitly by the equation itself, once the derivatives y' are known. If we differentiate (8), we obtain a second order equation involving only x, y', and y''. Such an equation can be regarded as a first order equation in y'. If this derived equation can be solved for y', then the solutions of (8) are obtained by substituting these values of y' in (8).

EXAMPLE 4. Consider again (cf. Example 1) the equation

$$y = xy' - \tfrac{1}{4}(y')^2. \tag{9}$$

Differentiation gives the equation

$$y' = xy'' + y' - \tfrac{1}{2}y'y'',$$
$$y''[x - \tfrac{1}{2}y'] = 0.$$

Therefore we must have $y' = c$ (if $y'' = 0$), or $y' = 2x$. Substitution in (9) gives all solutions,

$$y = cx - \tfrac{1}{4}c^2,$$

or

$$y = (2x)x - \tfrac{1}{4}(2x)^2 = x^2.$$

EXAMPLE 5.

$$y = xy' + \frac{2}{x} \tag{10}$$

The variables cannot be separated in (10), nor is it exact or a Clairaut equation. For any solution y, however, we must have

$$y' = xy'' + y' - \frac{2}{x^2},$$
$$y'' = 2x^{-3}.$$

Therefore y is a solution if and only if

$$y' = -x^{-2} + c,$$

and the solutions are, by substitution in (10),

$$y = x(-x^{-2} + c) + \frac{2}{x} = cx + \frac{1}{x}.$$

PROBLEMS

1. Find all solutions, including envelopes.

(a) $y = xy' + y'$

(b) $yy' = x(y')^2 + 1$

(c) $(y - xy')^2 + 2y' = 0$

(d) $y - xy' = \ln y'$

(e) $y = xy' + (y')^2$

2. Solve the following equations by first differentiating, and then finding all values of y'. (The solution for y' is frequently simpler if one substitutes u for y', and u' for y'' after differentiation.)

(a) $3y = xy' + 2x$

(b) $y = x + y' + xy'$

(c) $y = xy' - \ln |x|$

(d) $y = (x + 1)y' + x^2$

(e) $y' = 1 + x - y$

(f) $y + xy' = 2x + y'$

(g) $2yy' = 3x + x(y')^2$

(h) $y = y' - x + \ln |y'|$

3. (a) Show that the differential equation of the family $y = c \ln |x| + g(c)$ is $y = xy' \ln |x| + g(xy')$.

(b) Show, as in the proof of Theorem 1, that the only solutions of the differential equation of part (a) are the functions $y = c \ln |x| + g(c)$ and the envelopes of this family.

4. Solve the following equations (see Problem 3). Show that the envelopes of the families of solutions obtained by inspection can also be found by differentiating the equation as in Problem 2.

(a) $y = xy' \ln |x| - (xy')^2$

(b) $y = xy' \ln |x| + 1 + 1/xy'$

ANSWERS

1. (a) $y = cx + c$

(b) $y = cx + 1/c, \ y^2 = 4x$

(c) $(y - cx)^2 + 2c = 0, \ y = 1/2x$

(d) $y = cx + \ln c, \ y = -1 - \ln(-x)$

(e) $y = cx + c^2, \ y = -\frac{1}{4}x^2$

2. (a) $y = x + cx^3$

(b) $y = x + c(1 + x) - (x + 1) \ln |1 + x|$

(c) $y = cx - 1 - \ln |x|$

(d) $y = -x^2 - 2x + 2(x + 1) \ln |x + 1| + c(x + 1)$

(e) $y = x + ce^{-x}$

(f) $y = \dfrac{c}{1 - x} + x + 1$

(g) $2y = cx^2 + 3/c, \ y = \pm\sqrt{3}\,x$

(h) $y = ce^x + \ln |c|, \ y = -(1 + x)$

4. (a) $y = c \ln |x| - c^2, \ y = \frac{1}{4}(\ln |x|)^2$

(b) $y = c \ln |x| + 1 + 1/c, \ y = 2(\ln |x|)^{1/2} + 1$

CHAPTER 2

SPECIAL METHODS FOR FIRST ORDER EQUATIONS

2-1 Homogeneous equations—substitutions. We can change the form of a differential equation in the following way. If we let y be any solution of

$$y' = F(x, y) \tag{1}$$

and introduce a new (unknown) function u by an equation such as $y(x) = g(u(x))$, then u must satisfy the new equation

$$g'(u)u' = F(x, g(u)). \tag{2}$$

Conversely, if u satisfies (2), then y satisfies (1). This process of substitution is of course pointless unless the resulting equation is simpler than the original. We study here one type of equation for which a standard substitution, $y = xu$, always gives an equation in which the variables separate.

A function F of two variables is *homogeneous* (of degree zero) if and only if $F(tx, ty) = F(x, y)$ for all numbers x, y, t. The following are examples of homogeneous functions:

$$\frac{x + y}{x}, \quad \frac{x^3 + xy^2}{y^3 + x^2y}, \quad \frac{xy + x^2 \sin (y/x)}{y^2}. \tag{3}$$

If F is a homogeneous function, then $F(x, y)$ depends only on the ratio y/x, and hence F can be considered as a function of a single variable $u = y/x$. To see this, put $t = 1/x$ in the definition to obtain

$$F(x, y) = F\left(\frac{1}{x}\, x, \frac{1}{x}\, y\right) = F\left(1, \frac{y}{x}\right) = F(1, u).$$

The homogeneous functions (3), for example, can be written

$$\frac{x + y}{x} = 1 + \left(\frac{y}{x}\right),$$

$$\frac{x^3 + xy^2}{y^3 + x^2y} = \frac{1 + (y/x)^2}{(y/x)^3 + (y/x)},$$

$$\frac{xy + x^2 \sin (y/x)}{y^2} = \frac{(y/x) + \sin (y/x)}{(y/x)^2}.$$

A *homogeneous differential equation* is an equation of the form $y' = F(x, y)$, where F is homogeneous of degree zero. If we make the substitution $u = y/x$ in a homogeneous equation, we get, since $y' = u + xu'$ and $F(x, y) = F(1, y/x)$,

$$u + xu' = F(1, u). \tag{4}$$

The variables separate in (4) to give

$$\frac{u'}{F(1, u) - u} = \frac{1}{x}.$$

EXAMPLE 1.

$$y' = \frac{x + y}{x - y} = \frac{1 + (y/x)}{1 - (y/x)}.$$

We let $u = y/x$, hence $y' = u + xu'$, and the equation becomes

$$u + xu' = \frac{1 + u}{1 - u}.$$

Simplifying, and separating variables, we get

$$xu' = \frac{1 + u - u + u^2}{1 - u} = \frac{1 + u^2}{1 - u},$$

$$\frac{(1 - u)u'}{1 + u^2} = \frac{1}{x}.$$

Integration now yields

$$\tan^{-1}u - \tfrac{1}{2}\ln(1 + u^2) = \ln|x| + c,$$

and hence the solutions of the original equation are given by

$$\tan^{-1}\left(\frac{y}{x}\right) - \tfrac{1}{2}\ln\frac{x^2 + y^2}{x^2} = \ln|x| + c.$$

EXAMPLE 2. The equation

$$y' = \frac{x(y + 1) + (y + 1)^2}{x^2}$$

is not homogeneous, but becomes homogeneous after the substitution $v(x) = y(x) + 1$. In this case $v' = y'$, and v must satisfy

$$v' = \frac{xv + v^2}{x^2} = \frac{v}{x} + \left(\frac{v}{x}\right)^2.$$

Now let $u = v/x$, and obtain

$$xu' + u = u + u^2,$$
$$-u^{-1} = \ln |x| + c.$$

Reversing the two substitutions, we get

$$-\frac{x}{v} = \ln |x| + c,$$
$$-x = (y + 1) [\ln |x| + c].$$

EXAMPLE 3. The equation

$$y' = \frac{(x + 1)^2 y + y^3}{(x + 1)^3}$$

is not homogeneous, but becomes homogeneous if we consider the independent variable to be $x + 1$ instead of x. If we let $t = x + 1$ and write $dy/dt = dy/dx$, we get the homogeneous equation

$$\frac{dy}{dt} = \frac{t^2 y + y^3}{t^3} = \left(\frac{y}{t}\right) + \left(\frac{y}{t}\right)^3.$$

Now the substitution $u = y/t$, $dy/dt = t(du/dt) + u$, yields the equation

$$t \frac{du}{dt} + u = u + u^3,$$

which gives

$$-\tfrac{1}{2}u^{-2} = \ln |t| + c.$$

Finally, the solutions of the original equation are given by

$$-\frac{1}{2} \left(\frac{x + 1}{y}\right)^2 = \ln |x + 1| + c.$$

The mechanical process of substituting t for $x + 1$, and dy/dt for $dy/dx = y'$ can be justified by making a proper substitution. We illustrate the argument for the very simple equation

$$y' = \frac{y}{x + 1}. \tag{5}$$

If (5) is written out completely we have

$$y'(x) = \frac{y(x)}{x + 1}. \tag{6}$$

Since (6) is to be an identity for a solution y, the equation can be written equally well as

$$y'(x - 1) = \frac{y(x - 1)}{x}. \tag{7}$$

[For this paragraph only, "$y(x - 1)$" means "y of $x - 1$," etc.] In (7) we make the substitution

$$v(x) = y(x - 1), \quad v'(x) = y'(x - 1), \tag{8}$$

and (7) becomes

$$v'(x) = \frac{v(x)}{x}.$$

Separating variables and integrating, we get $v(x) = cx$, whence [by (8)] $y(x - 1) = cx$, and $y(x) = c(x + 1)$. We clearly end up with the same result with the formalism of Example 3. Let $t = x + 1$, $dy/dt = dy/dx$, in (5) to get

$$\frac{dy}{dt} = \frac{y}{t}.$$

Integration gives $y = ct$ and, by substitution, $y = c(x + 1)$.

The equation

$$y' = \frac{ax + by + c}{px + qy + r}, \tag{9}$$

with $aq - bp \neq 0$, can be made homogeneous by a substitution of the form

$$v(x) = y(x + \alpha) - \beta. \tag{10}$$

As indicated above, we can accomplish the same thing as the substitution (10) by writing

$$v = y - \beta, \quad t = x - \alpha, \quad \frac{dv}{dt} = \frac{dy}{dx}. \tag{11}$$

With the agreement (11), equation (9) becomes

$$\frac{dv}{dt} = \frac{at + bv + (a\alpha + b\beta + c)}{pt + qv + (p\alpha + q\beta + r)}. \tag{12}$$

If $aq - bp \neq 0$, we can find numbers α and β such that

$$a\alpha + b\beta + c = 0,$$
$$p\alpha + q\beta + r = 0.$$

For α and β which are solutions of these linear equations, equation (12) is homogeneous:

$$\frac{dv}{dt} = \frac{at + bv}{pt + qv} = \frac{a + b(v/t)}{p + q(v/t)}.$$

EXAMPLE 4.

$$y' = \frac{x + 2y - 4}{2x - y - 3}.$$

Let $t = x - \alpha$, $v = y - \beta$, $dv/dt = dy/dx$, and the equation becomes

$$v' = \frac{t + 2v + (\alpha + 2\beta - 4)}{2t - v + (2\alpha - \beta - 3)}.$$

The numbers α and β which satisfy

$$\alpha + 2\beta = 4, \qquad 2\alpha - \beta = 3$$

are $\alpha = 2$, $\beta = 1$. The substitution $t = x - 2$, $v = y - 1$ [properly, $v(x) = y(x + 2) - 1$] gives the equation

$$v' = \frac{t + 2v}{2t - v} = \frac{1 + 2(v/t)}{2 - (v/t)}.$$

Now make the substitution $u = v/t$, $v' = tu' + u$, to obtain

$$u + tu' = \frac{1 + 2u}{2 - u}.$$

Hence

$$tu' = \frac{1 + 2u - 2u + u^2}{2 - u},$$

$$\frac{(2 - u)u'}{1 + u^2} = \frac{1}{t}.$$

Integration gives

$$2 \tan^{-1} u - \tfrac{1}{2} \ln (1 + u^2) = \ln |t| + c,$$

and the solutions are given by

$$2 \tan^{-1} \left(\frac{v}{t}\right) - \tfrac{1}{2} \ln \left(1 + \frac{v^2}{t_2}\right) = \ln |t| + c,$$

$$2 \tan^{-1} \left(\frac{y - 1}{x - 2}\right) - \tfrac{1}{2} \ln \left[1 + \left(\frac{y - 1}{x - 2}\right)^2\right] = \ln |x - 2| + c.$$

PROBLEMS

Solve the differential equations in 1 through 5.

1. $y' = \dfrac{y + x}{x}$ 2. $y' = \dfrac{2y - x}{y}$ 3. $y' = \dfrac{y}{x + y}$

4. $y' = \dfrac{x^2 + y^2}{xy}$ 5. $y' = \dfrac{x}{x + y}$

6. Show that if $y' = F(x, y)$ is a homogeneous equation, and $y = g(x)$ is a solution, then $y = (1/a)g(ax)$ is a solution for any number a. Illustrate this fact with the solutions of Problem 1. Interpret the result geometrically.

7. Find an appropriate substitution and solve $(x - y)^2 y' = 1$.

8. Find a substitution which transforms $y^2 = 2yy'(x + 1) + 1$ into a Clairaut equation and solve it.

Solve the differential equations in 9 and 10.

9. $y' = \dfrac{2x - 5y + 3}{2x + 4y - 6}$ 10. $y' = \dfrac{x + y + 2}{2x + y - 1}$

11. The procedure in the text for problems of the form

$$y' = \frac{ax + by + c}{px + qy + r}$$

fails if $aq - bp = 0$. Show that in this case $ax + by = k(px + qy)$, and the substitution $u = px + qy$ separates the variables.

12. Solve $y' = \dfrac{2x + y + 1}{4x + 2y + 3}$.

ANSWERS

1. $y = x (\ln |x| + c)$ 2. $x = (y - x) [\ln |y - x| + c]$

3. $x = y (\ln |y| + c)$ 4. $y^2 = 2x^2 (\ln |x| + c)$

5. $\ln \left| 1 - \dfrac{y}{x} - \left(\dfrac{y}{x}\right)^2 \right| - \dfrac{1}{\sqrt{5}} \ln \left| \dfrac{\sqrt{5} + 1 + (2y/x)}{\sqrt{5} - 1 - (2y/x)} \right| = -2 \ln |x| + c$

7. $y = \frac{1}{2} \ln \left| \dfrac{x - y - 1}{x - y + 1} \right| + c$

8. $y^2 = cx + c + 1$

9. $(x - 1)^3 \left[4 \left(\dfrac{y - 1}{x - 1}\right) - 1 \right] \left[\dfrac{y - 1}{x - 1} + 2 \right]^2 = c$

10. $\ln \left| 1 - \dfrac{y + 5}{x - 3} - \left(\dfrac{y + 5}{x - 3}\right)^2 \right| + \dfrac{3}{\sqrt{5}} \ln \left| \dfrac{\dfrac{y + 5}{x - 3} + \dfrac{1}{2} - \dfrac{\sqrt{5}}{2}}{\dfrac{y + 5}{x - 3} + \dfrac{1}{2} + \dfrac{\sqrt{5}}{2}} \right|$

$$= -2 \ln |x - 3| + c$$

12. $10(2x + y) + \ln |5(2x + y) + 7| = 25x + c$

2–2 Exact equations. The first order equation

$$M(x, y)\, dx + N(x, y)\, dy = 0 \tag{1}$$

is called *exact* if there is a function u of two variables such that $u_x(x, y) = M(x, y)$ and $u_y(x, y) = N(x, y)$. That is, a function u such that

$$du(x, y) = M(x, y)\, dx + N(x, y)\, dy. \tag{2}$$

If there is a function u satisfying (2) for given functions M and N, then except for an arbitrary additive constant, there is exactly one such function. We saw in Section 1–2 that the solutions of (1), if (1) is exact, are just those functions y satisfying $u(x, y) = c$ for some constant c. In this section we find conditions on M and N which are necessary and sufficient that (1) be exact and methods for finding the function u.

THEOREM 1 (Necessary condition for exactness). *If* (1) *is exact and* M_y *and* N_x *are continuous, then* $M_y = N_x$. *Specifically, the equality* $M_y(x, y) = N_x(x, y)$ *must hold for all* (x, y) *in any region on which there is a function* u *satisfying the condition* (2).

Proof. Suppose that u is a function satisfying (2); i.e., that $u_x = M$ and $u_y = N$. The mixed second partial derivatives u_{xy} and u_{yx} are equal where they are continuous, so we must have

$$M_y(x, y) = u_{xy}(x, y) = u_{yx}(x, y) = N_x(x, y).$$

The necessary condition $M_y = N_x$ is also sufficient if some geometric restrictions are made about the region on which the condition holds. We show this next for the special case in which the region under consideration is a rectangle. The problem is treated in more generality in the next section.

THEOREM 2 (A sufficient condition for exactness). *If* M, N, M_y *and* N_x *are continuous, and* $M_y(x, y) = N_x(x, y)$ *for all* (x, y) *in some rectangle* R,

$$R = \{(x, y) : a \le x \le b \text{ and } c \le y \le d\},$$

then there is a function u *defined on* R *such that* $u_x(x, y) = M(x, y)$ *and* $u_y(x, y) = N(x, y)$ *for all* (x, y) *in* R.

Proof. We will show that the following function u satisfies the conditions of the theorem:

$$u(x, y) = \int_a^x M(t, c)\, dt + \int_c^y N(x, t)\, dt. \tag{3}$$

Note that $u(x, y)$ is defined for every (x, y) in R, since the integrands in

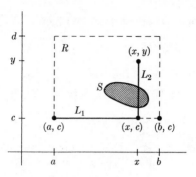

FIGURE 2–1

(3) are continuous (functions of t) on the intervals of integration (see Fig. 2–1). We will need the following formulas from calculus:

$$\frac{\partial}{\partial x} \int_a^x M(t, c) \, dt = M(x, c), \tag{4}$$

$$\frac{\partial}{\partial x} \int_c^y N(x, t) \, dt = \int_c^y N_x(x, t) \, dt, \tag{5}$$

$$\frac{\partial}{\partial y} \int_c^y N(x, t) \, dt = N(x, y). \tag{6}$$

Using (4) and (5), and the assumption that $M_y = N_x$ within the rectangle, we obtain* from (3)

$$
\begin{aligned}
u_x(x, y) &= M(x, c) + \int_c^y N_x(x, t) \, dt \\
&= M(x, c) + \int_c^y M_y(x, t) \, dt \\
&= M(x, c) + M(x, t) \Big]_{t=c}^{t=y} \\
&= M(x, c) + M(x, y) - M(x, c) \\
&= M(x, y). \tag{7}
\end{aligned}
$$

Since the first integral in (3) does not depend on y, we get immediately from (6)

$$u_y(x, c) = N(x, y). \tag{8}$$

---

* M_y is a function of two variables, and $M_y(x, t)$ is the value of this function at (x, t). Do not confuse $M_y(x, t)$ with $\partial/\partial y M(x, t)$, which is zero.

The first integral in (3) is the integral of M restricted to the line L_1 (Fig. 2-1) from (a, c) to (x, c), and the second is the integral of N restricted to the line L_2 from (x, c) to (x, y). We need to know that the functions are continuous, and the identity $M_y = N_x$ holds along the broken line path (L_1 and L_2) from (a, c) to any point (x, y) in R. If there were a region in R where the conditions of the theorem failed (e.g., the shaded region S of the figure), the argument above could not be used to show $du = M\,dx + N\,dy$, even for points not in S.

Rather than use the formula of the theorem, it is frequently simpler to deal with indefinite integrals. Formula (3) for u shows that if we integrate N with respect to y, we obtain a function which differs from u by a function of x only [namely, the first integral on the right of (3)].

Schematically,

$$u(x, y) = \int N(x, y)\,dy + \phi(x).$$

To find ϕ, we use the fact that we must have

$$u_x(x, y) = \frac{\partial}{\partial x} \int N(x, y)\,dy + \phi'(x) = M(x, y).$$

This gives a formula for $\phi'(x)$ from which ϕ can be found by integration. Similarly, we can write

$$u(x, y) = \int M(x, y)\,dx + \phi(y)$$

and find ϕ from the condition

$$\frac{\partial}{\partial y} \int M(x, y)\,dx + \phi'(y) = N(x, y).$$

EXAMPLE 1. $2xy\,dx + (x^2 + 2y)\,dy = 0$.

Since $(\partial/\partial y)(2xy) = 2x = (\partial/\partial x)(x^2 + 2y)$, the equation is exact. Hence

$$u(x, y) = \int 2xy\,dx + \phi(y)$$

$$= x^2 y + \phi(y).$$

The condition $u_y(x, y) = x^2 + 2y$ gives us

$$x^2 + \phi'(y) = x^2 + 2y, \qquad \phi'(y) = 2y.$$

Therefore, $u(x, y) = x^2 y + y^2$, and the solutions of the differential equation are the functions satisfying $x^2 y + y^2 = c$.

The formula (3), with $a = 2$, $c = 1$, applied to the above problem yields directly

$$u(x, y) = \int_2^x M(t, 1) \, dt + \int_1^y N(x, t) \, dt$$

$$= \int_2^x 2t \, dt + \int_1^y (x^2 + 2t) \, dt$$

$$= x^2 - 4 + x^2 y + y^2 - x^2 - 1$$

$$= x^2 y + y^2 - 5.$$

The functions defined by the equations $x^2 y + y^2 = c$ and $x^2 y + y^2 - 5 = c$ are of course the same.

Problems

Show that the following equations are exact and solve them.

1. $y \, dx + (x + 1) \, dy = 0$
2. $\sin y \, dx + (x \cos y + 1) \, dy = 0$
3. $(2xy^2 + 2) \, dx + 2x^2 y \, dy = 0$

4. $\dfrac{y}{x^2} \, dx = \dfrac{1}{x} \, dy$

5. $[e^x + y^2 \cos (xy)] \, dx + [\sin (xy) + xy \cos (xy)] \, dy = 0$
6. $(x^2 + y^2)(x \, dx + y \, dy) + 2 \, dx = 0$

7. $(\ln |y + 1| + y^2) \, dx + \left(\dfrac{x}{y + 1} + 2xy \right) dy = 0$

8. $y' = -\dfrac{2x + y}{x + 2y}$

9. With the hypotheses of Theorem 2, show that the integral of N along the segment from (a, c) to (a, y) plus the integral of M along the segment from (a, y) to (x, y) gives a function v such that

$$dv = M \, dx + N \, dy.$$

Prove that $v = u$. [*Hint:* The functions u and v differ by a constant. Show that $u = v$ at some point and hence that the constant is zero.]

10. If $du = M \, dx + N \, dy$, then u is also given by

$$u(x, y) = \int_0^1 [xM(tx, ty) + yN(tx, ty)] \, dt.$$

Use this integral to solve the following equations:

(a) $(1 + y) \, dx + x \, dy = 0$
(b) $(y^2 + 1) \, dx + 2xy \, dy = 0$
(c) $dx - \sin y \, dy = 0$

ANSWERS

1. $xy + y = c$ 2. $x \sin y + y = c$
3. $x^2y^2 + 2x = c$ 4. $x + y = cx$
5. $y \sin (xy) + e^x = c$ 6. $(x^2 + y^2)^2 + 8x = c$
7. $x \ln |y + 1| + xy^2 = c$ 8. $x^2 + xy + y^2 = c$
10. (a) $x + xy = c$ (b) $x + xy^2 = c$ (c) $x + \cos y - 1 = c$

2-3 Line integrals. The concept of an integral along a curve in the plane has many applications. We introduce the idea briefly here in order to clarify the treatment of exact equations.

A *curve* is a set of points of the form

$$C = \{(f(t), g(t)): a \leq t \leq b\}, \tag{1}$$

where f and g are functions with continuous derivatives on $[a, b]$. We will say that

$$x = f(t), \qquad y = g(t), \qquad (a \leq t \leq b) \tag{2}$$

are *parametric equations* for C. The curve C in (1) is *closed* if $(f(a), g(a)) = (f(b), g(b))$ and is a *simple closed curve* if $(f(t_1), g(t_1)) = (f(t_2), g(t_2))$ implies $t_1 = a$ and $t_2 = b$, or $t_1 = b$ and $t_2 = a$. That is, a simple closed curve does not intersect itself. (See Fig. 2-2.)

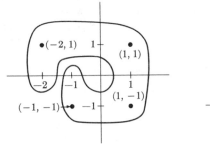

Simple closed curve

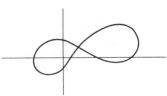

Nonsimple curve

FIGURE 2-2

If M and N are functions of two variables which are continuous on the curve C in (1), then we define the *line integral* of $M(x, y) \, dx + N(x, y) \, dy$ along C as

$$\int_C (M(x, y) \, dx + N(x, y) \, dy)$$
$$= \int_a^b [M(f(t), g(t))f'(t) + N(f(t), g(t))g'(t)] \, dt. \tag{3}$$

A curve C can be represented parametrically by infinitely many pairs of functions (f, g). The integral on the right side of (3), however, will give the same value for any pair of functions representing C if the curve is traced out in a given direction for increasing parameter values. The definition (3) therefore depends only on the set of points C, with a prescribed direction or orientation along C, and not on the particular choice of parametric functions for C.

The line integral (3) has the following physical interpretation. If M and N are respectively the horizontal and vertical components of a force field defined on the plane, then the integral (3) is the work done by this force field on a particle moving over the curve C in the prescribed direction.

As special cases $(N(x, y) = 0$ or $M(x, y) = 0)$ of (3), we have the formulas

$$\int_C M(x, y)\, dx = \int_a^b M(f(t),\ g(t))f'(t)\, dt, \tag{4}$$

$$\int_C N(x, y)\, dy = \int_a^b N(f(t), g(t))g'(t)\, dt. \tag{5}$$

EXAMPLE 1. The line segment C from $(0, 0)$ to $(1, 1)$ can be represented by the functions $f(t) = t$, $g(t) = t$, $0 \le t \le 1$, and equally well by the functions $h(t) = t^2 - 1$, $k(t) = t^2 - 1$, $1 \le t \le \sqrt{2}$. We compute $\int_C (M\, dx + N\, dy)$ for the functions $M(x, y) = x$, $N(x, y) = xy$, using both the above parameterizations, to illustrate that the integral depends only on C.

$$\int_C (x\, dx + xy\, dy) = \int_0^1 [t + t^2]\, dt = \tfrac{1}{2} + \tfrac{1}{3} = \tfrac{5}{6},$$

$$\int_C (x\, dx + xy\, dy) = \int_1^{\sqrt{2}} [(t^2 - 1)2t + (t^2 - 1)^2 2t]\, dt$$

$$= \int_1^{\sqrt{2}} (2t^5 - 2t^3)\, dt = \tfrac{1}{3}(8 - 1) - \tfrac{1}{2}(4 - 1) = \tfrac{5}{6}.$$

EXAMPLE 2. We compute $\int_C -y\, dx$, where C is the unit circle oriented in the counterclockwise direction. A convenient parametric representation is $x = f(t) = \cos t$, $y = g(t) = \sin t$, $0 \le t \le 2\pi$. We obtain

$$\int_C -y\, dx = \int_0^{2\pi} -\sin t\,(-\sin t)\, dt = \int_0^{2\pi} \sin^2 t\, dt = \pi.$$

EXAMPLE 3. We compute $\int_C x\, dy$, where C is the triangle consisting of the segments C_1: $(0, 0)$ to $(1, 0)$; C_2: $(1, 0)$ to $(0, 1)$; and C_3: $(0, 1)$ to $(0, 0)$.

The integral is the sum of the separate integrals taken over C_1, C_2, C_3. We have

$$\int_{C_1} x\, dy = 0,$$

$$\int_{C_2} x\, dy = \int_1^0 x(-1)\, dx = \tfrac{1}{2},$$

$$\int_{C_3} x\, dy = 0.$$

The first integral is zero since on C_1, y is constant, and $dy = 0$ (that is, the function g in (5) is constant, and $g'(t) \equiv 0$). The third integral is zero since $x = N(x, y) \equiv 0$ on C_3. For C_2 we used x as the parameter, with $y = -x + 1$, $dy = (-1)\, dx$, and x running from 1 to 0 to match the orientation of C_2. A more formal procedure would be $x = f(t) = 1 - t$, $y = g(t) = t$, $0 \le t \le 1$, giving

$$\int_{C_2} x\, dy = \int_0^1 (1 - t)\, dt = [t - \tfrac{1}{2}t^2]_0^1 = \tfrac{1}{2}.$$

It is not accidental (cf. Problem 7) that the line integrals of Examples 2 and 3 give the areas bounded by the closed curves.

A *domain* is a set G of points in the plane such that (i) G is *open*—each point of G is the center of a disc contained in G, and (ii) G is *connected*—any two points of G can be joined by a curve lying in G. The condition that G be open means that the set G does not contain any of its boundary points. For example, the set of points inside a simple closed curve, excluding the points of the curve itself, is a domain.

The next theorem shows that if $M\, dx + N\, dy$ is an exact differential, then the function u such that $du = M\, dx + N\, dy$ can be found as a line integral of $M\, dx + N\, dy$.

THEOREM 1. *If $M(x, y)\, dx + N(x, y)\, dy = du(x, y)$ for all (x, y) in a domain G, then*

$$\int_C (M(x, y)\, dx + N(x, y)\, dy) = u(c, d) - u(a, b)$$

for every curve C in G from (a, b) to (c, d).

Proof. Let C be any curve in G from (a, b) to (c, d), and let

$$x = f(t), \qquad y = g(t), \qquad \alpha \le t \le \beta$$

be parametric equations for C. In particular, $(f(\alpha), g(\alpha)) = (a, b)$, and $(f(\beta), g(\beta)) = (c, d)$. Define a function H on $[\alpha, \beta]$ by

$$H(t) = u(f(t), g(t)).$$

By our assumption that $u_x = M$, $u_y = N$, we have

$$H'(t) = u_x(f(t), g(t))f'(t) + u_y(f(t), g(t))g'(t)$$
$$= M(f(t), g(t))f'(t) + N(f(t), g(t))g'(t).$$

It follows that [cf. (3)]

$$\int_C (M(x, y)\, dx + N(x, y)\, dy) = \int_\alpha^\beta H'(t)\, dt$$
$$= H(\beta) - H(\alpha)$$
$$= u(f(\beta), g(\beta)) - u(f(\alpha), g(\alpha))$$
$$= u(c, d) - u(a, b).$$

For expressions $M\, dx + N\, dy$ which are exact differentials, we will write

$$\int_{(a,b)}^{(c,d)} (M(x, y)\, dx + N(x, y)\, dy)$$

to indicate the line integral taken over *any* curve from (a, b) to (c, d). By fixing a point (a, b), and taking any point (x, y) for (c, d) in (6), we get the following representation for a function u such that $du = M\, dx + N\, dy$.

$$u(x, y) = \int_{(a,b)}^{(x,y)} (M(x, y)\, dx + N(x, y)\, dy). \tag{6}$$

The formula given for u in Section 2–2 [Formula (3)] is just the line integral along a broken line path from (a, c) to (x, y).

We return now to the problem of determining to what extent the condition $M_y = N_x$ is sufficient that $M\, dx + N\, dy$ be exact. The relevant theorem depends on the following geometric notion. A domain G is *simply connected* if the region bounded by any simple closed curve in G contains only points of G. This says, roughly speaking, that G has no holes in it; i.e., no curve in G can surround points which are not in G. The following theorem is a consequence of Green's theorem (see Problem 6), and we will omit the proof.

THEOREM 2. *If $M_y(x, y) = N_x(x, y)$ for all (x, y) in a simply connected domain G, then there is some function u defined on G such that $du(x, y) = M(x, y)\, dx + N(x, y)\, dy$ for all (x, y) in G.*

Once we know that there is a function u defined on G, the function can be determined by the integral (6). The effect of picking different initial points (a, b) is simply to change the function u by an additive constant. The statements of Theorem 2, Section 2–2, are restricted to points within a rectangle. A rectangle is merely an example of a simply connected set, and moreover one in which every point can be reached with the simple broken line path used in that theorem. The formula of Problem 10, Section 2–2, is the line integral over the line segment from $(0, 0)$ to (x, y).

Problems

1. Compute the integrals for the curves specified.
   (a) $\int_C [(x + 1) \, dx - xy^2 \, dy]$; C is the parabola $y = x^2$ from $(-1, 1)$ to $(2, 4)$.
   (b) $\int_C [(x^2 + y) \, dx + (x + y) \, dy]$; C is the closed curve consisting of the arc of $y^2 = x$ from $(1, 1)$ to $(1, -1)$, and the segment from $(1, -1)$ to $(1, 1)$.
   (c) $\int_C [x^2 y \, dx + 1/(x^2 + 1) \, dy]$; C is the boundary of the unit square $\{(x, y) : 0 \le x \le 1, 0 \le y \le 1\}$ oriented counterclockwise.

2. Calculate $\int_C (x + y) \, dx$ for the two curves C_1 and C_2. C_1 is the segment from $(0, 0)$ to $(1, 1)$, and C_2 is the broken line from $(0, 0)$ to $(1, 0)$ and $(1, 0)$ to $(1, 1)$. Conclude that the hypothesis of Theorem 1 is necessary for a line integral to be independent of path.

3. Calculate $\int_{(1,2)}^{(x,y)} (2xy \, dx + x^2 \, dy)$, where the integral may be computed along any convenient curve.

4. (a) Is the domain G_0 consisting of all points of the plane except $(0, 0)$ simply connected?
   (b) Is the domain G consisting of all points of the plane except those of the form $(x, 0)$, $x \le 0$, simply connected?

5. Verify that the exactness condition of Theorem 2 holds for

$$\frac{-y}{x^2 + y^2} \, dx + \frac{x}{x^2 + y^2} \, dy$$

except at $(0, 0)$. Calculate

$$\int_C \frac{-y}{x^2 + y^2} \, dx + \frac{x}{x^2 + y^2} \, dy$$

for C_1: the upper half of the unit circle from $(1, 0)$ to $(-1, 0)$ and for C_2: the lower half circle from $(1, 0)$ to $(-1, 0)$. Reconcile this with Theorems 1 and 2.

6. Green's theorem states that if M, N, M_y, and N_x are continuous on the set G consisting of a simple closed curve C and its interior, and if C is oriented in the counterclockwise direction, then

$$\iint_G [N_x(x, y) - M_y(x, y)] \, dy \, dx = \int_C M(x, y) \, dx + N(x, y) \, dy.$$

Prove this for the case $N(x, y) \equiv 0$, when C is the closed curve formed by $y = g(x)$ and $y = f(x)$, with $g(a) = f(a)$, $g(b) = f(b)$, and $g(x) < f(x)$ for $a < x < b$.

7. Use Green's theorem (Problem 6) to show that $\int_C -y\,dx$ and $\int_C x\,dy$ give the area bounded by the simple closed counterclockwise curve C.

8. Show that $x = a\cos t$, $y = b\sin t$, $0 \le t \le 2\pi$, are parametric equations of the ellipse $x^2/a^2 + y^2/b^2 = 1$. Use either formula of Problem 7 to find the area bounded by this ellipse.

9. Solve the differential equations and specify the domains on which the solutions hold.

(a) $\dfrac{1}{\sqrt{x-y}}\,dx - \dfrac{1}{\sqrt{x-y}}\,dy = 0$ (b) $\left(2x + \dfrac{y}{x}\right)dx + \ln x\,dy = 0$

(c) $\left(\tan^{-1}\dfrac{y}{x} - \dfrac{xy}{x^2 + y^2}\right)dx + \dfrac{x^2}{x^2 + y^2}\,dy = 0$

10. The differential of Problem 5 is exact in the simply connected domain bounded by the simple closed curve of Fig. 2-2. There is, therefore, a function u defined on this domain such that $u(1, 0) = 0$ and $du(x, y) = -y/(x^2 + y^2)\,dx + x/(x^2 + y^2)\,dy$. Find $u(-1, 0)$ and $u(-2, 0)$.

ANSWERS

1. (a) $-\frac{453}{14}$ 2. $1, \frac{1}{2}$ 3. $x^2y - 2$
  (b) 0
  (c) $-\frac{5}{6}$

4. No, yes   5. $\pi, -\pi$ 8. πab

9. (a) $x - y = c^2$ for points below $y = x$
  (b) $x^2 + y\ln|x| = c$ for the right half-plane, or the left half-plane
  (c) $x\tan^{-1}(y/x) = c$ for the right half-plane, or the left half-plane

10. $u(-1, 0) = -\pi$, $u(-2, 0) = \pi$

2–4 First order linear equations — integrating factors.

A *linear* equation is any equation of the form

$$y^{(n)} + p_{n-1}(x)y^{(n-1)} + \cdots + p_1(x)y' + p_0(x)y = q(x).$$

We will study linear equations in detail in Chapters 3 and 4. Here we treat the simplest case, the first order linear equation

$$y' + p(x)y = q(x). \tag{1}$$

We assume that the functions p and q are continuous on some interval $[a, b]$. With this assumption it follows from Theorem 1, Section 1–4, that there is a solution of (1) through each point (x_0, y_0) such that $a < x_0 < b$. Although we cannot conclude from this theorem that the solutions of (1) are defined on the whole interval $[a, b]$, this is the case, as we show below.

First consider the special case of (1) in which $q(x) \equiv 0$,

$$y' + p(x)y = 0. \tag{2}$$

In (2) we can separate the variables and solve as follows:

$$\frac{y'}{y} = -p(x), \quad \text{or} \quad y = 0;$$

$$\ln |y| = -\int p(x)\, dx + \ln c \quad (c > 0), \quad \text{or} \quad y = 0;$$

$$|y| = c e^{-\int p(x)\, dx} \quad (c > 0), \quad \text{or} \quad y = 0;$$

$$y = c e^{-\int p(x)\, dx} \quad (c \text{ arbitrary}).$$

If we write the solutions in the form

$$y e^{\int p(x)\, dx} = c,$$

then differentiation gives

$$y' e^{\int p(x)\, dx} + p(x) y e^{\int p(x)\, dx} = 0.$$

That is, Eq. (2) becomes exact if multiplied by

$$e^{\int p(x)\, dx}. \tag{3}$$

If a differential equation becomes exact after multiplication by some function, such as (3), this function is called an *integrating factor* for the equation.

Now return to equation (1), and multiply both sides of this equation by the function (3). We get the equivalent equation

$$y' e^{\int p(x)\, dx} + p(x) y e^{\int p(x)\, dx} = e^{\int p(x)\, dx} q(x). \tag{4}$$

The left side of (4) is an exact differential, as we have seen. Since the right side of (4) does not involve y, it also is an exact differential, and therefore $e^{\int p(x)\, dx}$ is an integrating factor for (1) as well as (2). The solutions of (4), and hence (1), can be written

$$y e^{\int p(x)\, dx} = \int e^{\int p(x)\, dx} q(x)\, dx + c,$$

or, in explicit form,

$$y = e^{-\int p(x)\, dx} \left[\int e^{\int p(x)\, dx} q(x)\, dx + c \right]. \tag{5}$$

We assemble these facts in the following statement.

THEOREM 1. *If p and q are continuous on $[a, b]$, and $P'(x) = p(x)$ for all x in $[a, b]$, then $e^{P(x)}$ is an integrating factor for* (1). *The solutions of* (1) *are defined on all of $[a, b]$, and are the functions*

$$y(x) = e^{-P(x)} \left[\int_a^x e^{P(t)} q(t) \, dt + c \right].$$

EXAMPLE 1. $y' + 2xy = 3x$.

We can write the solutions directly from Formula (5).

$$\begin{aligned}
y &= e^{-\int 2x \, dx} \left[\int e^{\int 2x \, dx} 3x \, dx + c \right] \\
&= e^{-x^2} \left[\int e^{x^2} 3x \, dx + c \right] \\
&= e^{-x^2} [\tfrac{3}{2} e^{x^2} + c] \\
&= \tfrac{3}{2} + ce^{-x^2}.
\end{aligned}$$

Alternatively, we find the integrating factor $e^{\int 2x \, dx} = e^{x^2}$ and write

$$e^{x^2} y' + y e^{x^2} 2x = e^{x^2} 3x.$$

Integration gives

$$e^{x^2} y = \int e^{x^2} 3x \, dx = \tfrac{3}{2} e^{x^2} + c.$$

EXAMPLE 2.

$$y' - \frac{1}{x} y = -\frac{2}{x^2}.$$

The integrating factor is

$$e^{-\int (1/x) \, dx} = e^{-\ln x} = (e^{\ln x})^{-1} = \frac{1}{x}.$$

Multiplying by $1/x$, we get

$$\frac{1}{x} y' - \frac{1}{x^2} y = -\frac{2}{x^3},$$

and hence the solutions are given by

$$\frac{1}{x} y = \frac{1}{x^2} + c,$$

$$y = \frac{1}{x} + cx.$$

Now we look at an example of an integrating factor for a nonlinear equation.

EXAMPLE 3. $y\,dx + (2xy^2 + x)\,dy = 0$.

This equation is not exact ($M_y = 1 \neq N_x = 2y^2 + 1$) and is not linear, since y^2 appears. We write the equation in the form

$$x\,dy + y\,dx + 2xy^2\,dy = 0.$$

The part $x\,dy + y\,dx = d(xy)$ will remain exact if multiplied by any function of xy. The remaining term, $2xy^2\,dy$, becomes an exact differential if multiplied by $(xy)^{-1}$, which is therefore an integrating factor. The equation becomes

$$(xy)^{-1}(x\,dy + y\,dx) + 2y\,dy = 0,$$

or

$$d\,[\ln|xy| + y^2] = 0. \tag{6}$$

Notice that $y = 0$ is a solution of the original equation, but not of (6). The solutions therefore satisfy

$$y = 0, \qquad \text{or} \qquad \ln|xy| + y^2 = c,$$

which can be expressed in the single formula

$$xye^{y^2} = c.$$

Finding integrating factors is generally just a matter of recognizing the common differentials. However, we can determine whether a first order equation has an integrating factor which depends only on x and give a formula for the integrating factor in this case.

Suppose the equation

$$M(x, y)\,dx + N(x, y)\,dy = 0 \tag{7}$$

has the integrating factor $\mu(x)$; that is, suppose the following equation is exact.

$$\mu(x)M(x, y)\,dx + \mu(x)N(x, y)\,dy = 0. \tag{8}$$

By the results of Section 2–2, we must have

$$\frac{\partial}{\partial y}\,[\mu(x)M(x, y)] = \frac{\partial}{\partial x}\,[\mu(x)N(x, y)],$$

or

$$\mu(x)M_y(x, y) = \mu'(x)N(x, y) + \mu(x)N_x(x, y).$$

Rearranging this, we get

$$\frac{\mu'(x)}{\mu(x)} = \frac{M_y(x, y) - N_x(x, y)}{N(x, y)}. \tag{9}$$

Hence (9) is a necessary and sufficient condition that (8) be exact. Since the left side of (9) depends only on x, it follows that the right side of (9) must be a function of x only.

$$\frac{M_y(x, y) - N_x(x, y)}{N(x, y)} = \phi(x).$$

Hence μ must satisfy

$$\frac{\mu'(x)}{\mu(x)} = \phi(x),$$

from which we get

$$\mu(x) = e^{\int \phi(x)\, dx}.$$

We have proved the following theorem.

THEOREM 2. *If* $[M_y(x, y) - N_x(x, y)]/N(x, y) = \phi(x)$, *a function of* x *only, then* $e^{\int \phi(x)\, dx}$ *is an integrating factor for Eq. (7).*

EXAMPLE 4. $(y^2 + 1 + x)\, dx + 2y\, dy = 0.$

The condition of Theorem 2 is satisfied, since

$$\frac{M_y - N_x}{N} = \frac{2y - 0}{2y} = 1,$$

which does not involve y. Therefore $e^{\int 1\, dx} = e^x$ is an integrating factor, and

$$[e^x y^2 + (1 + x)e^x]\, dx + 2y e^x\, dy = 0$$

is exact. The solutions are

$$e^x y^2 + x e^x = c.$$

PROBLEMS

1. Solve the following equations:

(a) $y' + (1/x)y = 4x^2$ (b) $y' + y = e^x$
(c) $y' + 2xy = 2x^3$ (d) $(x - 1)y' + xy = e^{-x}$
(e) $y' + (\cot x)y = 3 \sin x \cos x$ (f) $y'' + 2y' = 4x$

2. Show that $1/x^2$, $1/y^2$, and $1/(x^2 + y^2)$ are integrating factors for $x\,dy - y\,dx = 0$.

3. Find an integrating factor and solve.

(a) $2y\,dx + x\,dy = 0$
(b) $2xy^2\,dx + 3x^2y\,dy = 0$
(c) $(x - y)\,dx + x\,dy = 0$
(d) $(2y^3 + 6xy^2)\,dx + (3xy^2 + 4x^2y)\,dy = 0$
(e) $(x \ln |xy| + x)\,dx + (x^2/y)\,dy = 0$
(f) $(\sqrt{x} + y)\,dx + 2x\,dy = 0$

4. An equation of the form

$$y' + p(x)y = q(x)y^n,$$

or $(m = -n)$,

$$y^m y' + p(x)y^{m+1} = q(x),$$

is called a *Bernoulli* equation. Find a substitution which transforms a Bernoulli equation into a linear equation in a new unknown.

5. Solve the following equations:

(a) $yy' - (2/x)y^2 = 1$
(b) $y' + y = 3e^x y^3$

6. Show directly from Theorem 1, Section 1–4, that if y is a solution of (2) and $y(x_0) = 0$ for some x_0, then $y(x) \equiv 0$.

7. Assume y_0 is a nonzero solution of (2), and y_1 is a solution of (1).

(a) Show by substitution that $y_1 + cy_0$ is a solution of (1) for every number c.
(b) Show by Theorem 1, Section 1–4, that every solution y of (1) can be written $y = y_1 + cy_0$ for some c.

8. Assume that y_1 and y_2 are distinct solutions of (1), and let $y_0 = y_1 - y_2$.

(a) Show that y_0 is a solution of (2).
(b) Write all solutions of (1) in terms of y_1 and y_2 (cf. Problem 7).

9. Check that the functions below are solutions of the given linear equation, and write all solutions of each equation by the method of Problem 8.

(a) $1 - e^{-x}$, $1 + 2e^{-x}$; $y' + y = 1$
(b) $x - 1/x$, $x + 1/x$; $y' + (1/x)y = 2$
(c) x, x^2; $(x^2 - x)y' + (1 - 2x)y = -x^2$

10. Show that if y_0 is a nonzero solution of (2), then (1) can be solved by the substitution $y(x) = y_0(x)v(x)$, and the solutions are $y = y_0[\int (q/y_0)\,dx + c]$. (This method, called "variation of parameters," extends to higher order linear equations.)

11. Solve the equations of Problem 9 by the variation of parameters method (Problem 10).

ANSWERS

1. (a) $y = x^3 + c/x$
   (b) $y = \frac{1}{2}e^x + ce^{-x}$
   (c) $y = x^2 - 1 + ce^{-x^2}$
   (d) $e^x(x - 1)y = x + c$
   (e) $y = \sin^2 x + c \csc x$
   (f) $y = x^2 - x + c_1 e^{-2x} + c_2$

3. (a) $x^2 y = c$
   (b) $x^2 y^3 = c$
   (c) $y/x + \ln |x| = c$
   (d) $x^2 y^3 + 2x^3 y^2 = c$
   (e) $x \ln |xy| = c$
   (f) $x + 2y\sqrt{x} = c$

5. (a) $y^2 = -\frac{2}{3}x + cx^4$
   (b) $y^2(6e^x + ce^{2x}) = 1, \ y = 0$

8. (b) $y = y_1 + c(y_1 - y_2)$

9. (a) $y = 1 + ce^{-x}$
   (b) $y = x + c/x$
   (c) $y = x + c(x^2 - x)$

2–5 Orthogonal families. We will say that two curves are *orthogonal* if and only if their tangents at points of intersection are perpendicular lines. Two *families* of curves are orthogonal if and only if every curve of the first family is orthogonal to every curve of the second family, and *vice versa*. Given a family of curves, the *orthogonal family*, or family of *orthogonal trajectories*, is the set of all curves which are orthogonal to every curve in the given family.

Orthogonal families of curves occur frequently in physical situations. For example, the lines of force in a two-dimensional force field are orthogonal to the equipotential curves. In the study of heat conduction, one finds that the isothermal curves are the orthogonal trajectories of the lines of heat flow.

If

$$y' = F(x, y)$$

is the differential equation of a family of curves, then the line tangent to one of these curves at a point (x, y) has slope $F(x, y)$. For an orthogonal trajectory of the family, the tangent line at (x, y) must have slope $-1/F(x, y)$, since lines are perpendicular if and only if their slopes are negative reciprocals. Hence the differential equation of the family of orthogonal trajectories is

$$y' = -\frac{1}{F(x, y)} \cdot$$

There are two steps in finding the family orthogonal to a given one-parameter family of curves. First, find the differential equation of the family by differentiating and eliminating the constant. We repeat: *eliminate the constant*. Second, replace y' by $-1/y'$ to obtain the differential equation of the orthogonal family and solve this differential equation.

EXAMPLE 1. Find the family orthogonal to $y = cx^2$.

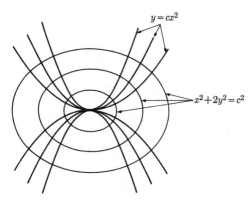

FIGURE 2-3

Differentiation gives $y' = 2cx$, or $c = y'/2x$. Substituting this value of c in the original equation, we obtain the differential equation of the family,

$$y = \tfrac{1}{2}xy'.$$

The differential equation of the orthogonal family is therefore

$$y = \frac{1}{2}x\left(-\frac{1}{y'}\right),$$

or

$$4yy' + 2x = 0.$$

The orthogonal family is the family of ellipses (see Fig. 2-3),

$$2y^2 + x^2 = c^2.$$

EXAMPLE 2. Show that the family $y^2 = 2cx + c^2$ is self-orthogonal.

Differentiating, we get $2yy' = 2c$, or $c = yy'$. Therefore, the differential equation of this family is

$$y^2 = 2xyy' + (yy')^2.$$

Substituting $-1/y'$ for y' we obtain the differential equation of the orthogonal family

$$y^2 = 2xy\left(-\frac{1}{y'}\right) + \left(\frac{y}{y'}\right)^2,$$

or

$$(yy')^2 = -2xyy' + y^2.$$

Since this last equation is the same as that for the given family, the family is self-orthogonal.

PROBLEMS

Find the orthogonal families and sketch both families in Problems 1 through 5.

1. $x^2 + y^2 = c^2$ 2. $y = cx^4$ 3. $(x - c)^2 + y^2 = c^2$

4. $y = \dfrac{c}{1 + x^2}$ 5. $y = ce^x$

6. Sketch the self-orthogonal family of curves of Example 2.

7. Show that if every curve in the family of rectangular hyperbolas $xy = c$ is rotated $45°$, the resulting family is orthogonal to the original.

8. Show that $x^2/c + y^2/(c - \alpha^2) = 1$ is a family of ellipses and hyperbolas with foci at $(\pm\alpha, 0)$. Show that the differential equation of the family is $(x + yy')(x - y/y') = \alpha^2$, and hence that the family is self-orthogonal.

ANSWERS

1. $y = cx$ 2. $x^2 + 4y^2 = c^2$ 3. $x^2 + (y - c)^2 = c^2$

4. $y^2 = \ln|x| + \frac{1}{2}x^2 + c$ 5. $y^2 = -2x + c$

2–6 Review of power series.

An indicated infinite sum of the form

$$\sum_{n=0}^{\infty} a_n x^n = a_0 + a_1 x + a_2 x^2 + \cdots, \tag{1}$$

or of the form

$$\sum_{n=0}^{\infty} a_n (x - c)^n = a_0 + a_1(x - c) + a_2(x - c)^2 + \cdots \tag{2}$$

is called a power series. We recall a few of the basic facts about power series.

The series (1) *converges at* x_0 *to* $f(x_0)$ if and only if

$$\lim_{N \to \infty} \sum_{n=0}^{N} a_n x_0^n = f(x_0). \tag{3}$$

If the limit (3) fails to exist, the series (1) *diverges at* x_0. If (1) converges at x_0 and $|x_1| < |x_0|$, then (1) converges at x_1. If (1) diverges at x_0 and $|x_1| > |x_0|$, then (1) diverges at x_1. It follows that (1) converges only at zero, or converges everywhere, or there is a positive number r such that (1) converges on $(-r, r)$ and diverges outside $[-r, r]$. The number r is called the *radius of convergence*, and the interval $(-r, r)$ is called the *interval of convergence* of (1). A series may or may not converge at either endpoint of its interval of convergence. For a series of the form (2), the interval of convergence is an interval of the form $(c - r, c + r)$.

Power series *converge absolutely* on their intervals of convergence; that is

$$\sum_{n=0}^{\infty} |a_n x^n| \quad \text{and} \quad \sum_{n=0}^{\infty} |a_n(x - c)^n|$$

converge on the intervals of convergence of

$$\sum_{n=0}^{\infty} a_n x^n \quad \text{and} \quad \sum_{n=0}^{\infty} a_n(x - c)^n.$$

This insures that we can rearrange the terms of a power series without affecting the convergence, or the value of the sum. Another consequence of absolute convergence is that one can multiply two power series by writing down in any order all products containing one term from each series.

Power series *converge uniformly* on any *closed* interval in the interval of convergence, but not in general on the whole interval of convergence. That is, if $(-r, r)$ is the interval of convergence of

$$\sum_{n=0}^{\infty} a_n x^n = f(x),$$

and $-r < a < b < r$, then for every $\epsilon > 0$, there is some N_0 not depending on x, such that

$$\left| f(x) - \sum_{n=0}^{N} a_n x^n \right| < \epsilon$$

for all x in $[a, b]$, whenever $N \geq N_0$. In other words, the polynomial which is the Nth partial sum of the series will be uniformly close to $f(x)$ on $[a, b]$, provided N is sufficiently large. If the series converges on the whole line, then it converges uniformly on any finite interval $[a, b]$.

If $f(x) = \sum_{n=0}^{\infty} a_n x^n$ for x in the interval of convergence, $(-r, r)$, then f is continuous and has derivatives of all orders on $(-r, r)$. The derivatives of f are given by the series obtained by termwise differentiation of the series for f. That is, termwise differentiation does not change the interval of convergence of a series, and if

$$f(x) = \sum_{n=0}^{\infty} a_n x^n, \qquad x \text{ in } (-r, r), \tag{4}$$

then

$$f'(x) = \sum_{n=1}^{\infty} n a_n x^{n-1}, \qquad x \text{ in } (-r, r). \tag{5}$$

A power series may be integrated term by term over any closed interval within the interval of convergence. Thus if f is given by (4), r is the radius of convergence of the series, and $-r < a < b < r$, then $\int_a^b f(x)\,dx$ exists since f is continuous on $[a, b]$. Moreover, the series (of constants)

$$\sum_{n=0}^{\infty} a_n \int_a^b x^n\,dx = \sum_{n=0}^{\infty} \frac{a_n}{n+1}\,(b^{n+1} - a^{n+1})$$

will converge, and the following identity holds:

$$\int_a^b f(x)\,dx = \sum_{n=0}^{\infty} a_n \int_a^b x^n\,dx.$$

By integrating over the interval $[0, x]$, or $[x, 0]$, for $|x| < r$, we can obtain indefinite integrals (antiderivatives) of f:

$$\int_0^x f(t)\,dt = \sum_{n=0}^{\infty} \frac{a_n}{n+1}\,x^{n+1}, \qquad -r < x < r.$$

The series on the right above, obtained by termwise integration of the series for f, will have the same radius of convergence as the series for f.

We will say that f is (real) *analytic* on an interval (a, b) if there is a series of the form (2) which converges to f on some interval around c, for each c in (a, b). Any function given by a convergent power series is analytic on the interval of convergence of the series. Thus if $f(x) = \sum a_n x^n$ for $|x| < r$, and $|c| < r$, there is a series $\sum b_n (x - c)^n$ which converges to $f(x)$ on some interval around c.

The common functions e^x, $\sin x$, and $\cos x$ are analytic on the whole line; their series at zero converge for all x:

$$e^x = \sum_{n=0}^{\infty} \frac{1}{n!}\,x^n, \tag{6}$$

$$\sin x = \sum_{n=1}^{\infty} (-1)^{n+1} \frac{1}{(2n-1)!}\,x^{2n-1}, \tag{7}$$

$$\cos x = \sum_{n=0}^{\infty} (-1)^n \frac{1}{(2n!)}\,x^{2n}. \tag{8}$$

The function $\sqrt{x}$ is analytic on $(0, \infty)$. If $c > 0$, there is a series for $\sqrt{x}$ of the form $\sum_{n=0}^{\infty} a_n (x - c)^n$ which will converge on $(0, 2c)$. The function $1/(1 + x^2)$ is analytic on the whole line. The series for $1/(1 + x^2)$ at zero is the geometric series $1 - x^2 + x^4 - x^6 + \cdots$ which converges on

$(-1, 1)$. The series at the point c will have radius of convergence $\sqrt{1 + c^2}$ (the distance between the complex numbers c and $i = \sqrt{-1}$).

The simplest way of determining the radius of convergence is the *ratio test*. If

$$\lim_{n \to \infty} \left| \frac{a_n}{a_{n+1}} \right| = r, \tag{9}$$

then r is the radius of convergence of (1) or (2). If

$$\lim_{n \to \infty} \left| \frac{a_{n+1}}{a_n} \right| = 0, \tag{10}$$

[note that the ratio in (10) is the reciprocal of that in (9)], then the series converges for all x.

If the ratio test fails because $\{a_n/a_{n+1}\}$ fails to converge, one can frequently use the following *comparison test*. If $|a_n| \leq |b_n|$ for all n, and $\sum b_n x^n$ has radius of convergence r, then $\sum a_n x^n$ has radius of convergence at least r. For example, the ratio test fails for the sine series (7) since every other coefficient is zero, but the series can be shown to converge everywhere by comparison with the exponential series (6).

If a function f is analytic on an interval around c, then the coefficients in the series for f at c are

$$a_n = \frac{1}{n!} f^{(n)}(c), \qquad n = 0, 1, 2, \ldots, \tag{11}$$

and hence the series is

$$f(x) = \sum_{n=0}^{\infty} \frac{1}{n!} f^{(n)}(c)(x - c)^n.$$

Problems

1. Find the interval of convergence for

(a) $\displaystyle\sum_{n=0}^{\infty} n^2 x^n$

(b) $\displaystyle\sum_{n=0}^{\infty} n! x^n$

(c) $\displaystyle\sum_{n=0}^{\infty} \frac{1}{2^n} x^n$

(d) $1 + 2x + x^2 + 2x^3 + x^4 + 2x^5 + \cdots$.

2. Find the series and its interval of convergence for the indicated function about the given point.

(a) $\dfrac{1}{1 - x}$, $c = 0$

(b) $\sqrt{x}$, $c = 1$

(c) $\ln x$, $c = 1$

(d) $\ln |x|$, $c = -1$

3. Use the geometric series Problem 2(a) to find series for these functions. Give the interval of convergence in each case.

(a) $\dfrac{1}{1 - x^2}$ 　　　 (b) $\dfrac{1}{1 + 2x}$ 　　　 (c) $\ln (1 + 2x)$

(d) $\dfrac{2x}{(1 - x^2)^2}$ 　　 (e) $\dfrac{1}{2 + x}$ 　　　 (f) $\tan^{-1} x$

4. Use (6), (7), and (8) to find series for

(a) e^{x^2} 　　 (b) $\sin (2x)$ 　　 (c) $\cos \sqrt{x}$ 　　 (d) $\sin (x - 1)$.

5. Derive formula (11). [*Hint:* If $f(x)$ is given by (4), then $f(0) = a_0$. Since $f'(x)$ is given by (5), $f'(0) = a_1$, etc.]

6. Suppose $\sum a_n x^n$ converges if and only if $-1 \le x < 1$. For what values of x do the following converge?

(a) $\sum a_n (x - 2)^n$ 　　 (b) $\sum (-1)^n a_n (x + 3)^n$ 　　 (c) $\sum a_n x^{2n}$

ANSWERS

1. (a) $(-1, 1)$ 　　 (b) $x = 0$ 　　 (c) $(-2, 2)$ 　　 (d) $(-1, 1)$

2. (a) $\displaystyle\sum_{n=0}^{\infty} x^n$,　$(-1, 1)$

(b) $1 + \displaystyle\sum_{n=1}^{\infty} (-1)^{n+1} \dfrac{1 \cdot 3 \cdot 5 \cdots (2n - 1)}{n! 2^n} (x - 1)^n$,　$(0, 2)$

(c) $\displaystyle\sum_{n=1}^{\infty} \dfrac{(-1)^{n+1}}{n} (x - 1)^n$,　$(0, 2)$

(d) $\displaystyle\sum_{n=1}^{\infty} \dfrac{-1}{n} (x + 1)^n$,　$(-2, 0)$

3. (a) $\displaystyle\sum_{n=0}^{\infty} x^{2n}$,　$(-1, 1)$ 　　　 (b) $\displaystyle\sum_{n=0}^{\infty} (-2)^n x^n$,　$(-\tfrac{1}{2}, \tfrac{1}{2})$

(c) $\displaystyle\sum_{n=0}^{\infty} \dfrac{(-2)^n}{n + 1} x^{n+1}$,　$(-\tfrac{1}{2}, \tfrac{1}{2})$ 　 (d) $\displaystyle\sum_{n=0}^{\infty} 2n x^{2n-1}$,　$(-1, 1)$

(e) $\displaystyle\sum_{n=0}^{\infty} \dfrac{(-1)^n}{2^{n+1}} x^n$,　$(-2, 2)$ 　 (f) $\displaystyle\sum_{n=0}^{\infty} \dfrac{(-1)^n}{2n + 1} x^{2n+1}$,　$(-1, 1)$

4. (a) $\displaystyle\sum_{n=0}^{\infty} \dfrac{x^{2n}}{n!}$ 　　　 (b) $\displaystyle\sum_{n=1}^{\infty} (-1)^{n+1} \dfrac{2^{2n-1}}{(2n - 1)!} x^{2n-1}$

(c) $\displaystyle\sum_{n=0}^{\infty} (-1)^n \dfrac{1}{(2n)!} x^n$ 　 (d) $\displaystyle\sum_{n=1}^{\infty} (-1)^{n+1} \dfrac{1}{(2n - 1)!} (x - 1)^{2n-1}$

6. (a) $1 \le x < 3$ 　　 (b) $-4 < x \le -2$ 　　 (c) $-1 < x < 1$

2–7 Series solutions. Even the simplest type of differential equation may have solutions which cannot be written in terms of elementary functions, such as polynomials, exponentials, trigonometric functions, etc. For example, the solutions of

$$y' = e^{-x^2} \tag{1}$$

are simply the antiderivatives of e^{-x^2}, but we cannot express these with a simple formula. We can, however, write a power series for each solution of (1). Since

$$e^x = 1 + x + \frac{1}{2!} x^2 + \cdots + \frac{1}{n!} x^n + \cdots$$

for all x, it follows that

$$e^{-x^2} = 1 - x^2 + \frac{1}{2!} x^4 - \cdots + \frac{1}{n!} (-x^2)^n + \cdots, \tag{2}$$

and hence the solutions of (1) are given by

$$c + \int_0^x e^{-t^2} \, dt = c + x - \tfrac{1}{3}x^3 + \frac{1}{5 \cdot 2!} x^5 - \cdots.$$

In general we do not know whether the solutions of a given differential equation are analytic, so we proceed as follows. We *suppose* that a differential equation has a series solution,

$$y = a_0 + a_1 x + a_2 x^2 + \cdots \tag{3}$$

and substitute the series in the equation to determine what the coefficients must be. That is, we find conditions on $a_0, a_1, a_2, \ldots$, which are necessary for (3) to be a solution of the given equation. Then we must check whether the series found in this way converges, and whether it is in fact a solution.

EXAMPLE 1. Suppose that

$$y' + 2xy = 1 \tag{4}$$

has a solution y which is analytic on some interval around zero. Thus

$$y = a_0 + a_1 x + a_2 x^2 + \cdots + a_n x^n + \cdots. \tag{5}$$

We know that y' is given on the same interval by the series

$$y' = a_1 + 2a_2 x + 3a_3 x^2 + \cdots + na_n x^{n-1} + \cdots. \tag{6}$$

Therefore (5) is a solution of (4) if and only if

$$a_1 + 2a_2x + \cdots + na_nx^{n-1} + \cdots$$
$$+ 2x(a_0 + a_1x + \cdots + na_nx^n + \cdots) = 1 \qquad (7)$$

is an identity on some interval around zero. Collecting terms, we see that (7) is equivalent to

$$a_1 + (2a_0 + 2a_2)x + (2a_1 + 3a_2)x^2 + \cdots$$
$$+ (2a_{n-2} + na_n)x^{n-1} + \cdots = 1. \qquad (8)$$

Since two power series converge to the same function if and only if the corresponding coefficients are equal [by (11), Section 2–6], and the right side of (8) can be regarded as the series with coefficients $1, 0, 0, 0, \ldots$, it follows that (8) is equivalent to the relations

$$\begin{aligned}
a_1 &= 1, \\
2a_0 + 2a_2 &= 0, \\
2a_1 + 3a_3 &= 0, \\
&\vdots \\
2a_{n-2} + na_n &= 0.
\end{aligned} \qquad (9)$$

The recursion relation

$$a_n = -\frac{2}{n} a_{n-2}, \qquad (n \geq 2), \qquad (10)$$

determines a_n, for n odd, in terms of $a_1 = 1$. Thus $a_3 = -\frac{2}{3}$, $a_5 = \frac{4}{15}$, and in general

$$\begin{aligned}
a_{2n+1} &= \frac{(-1)^n 2^n}{(2n+1)(2n-1)\cdots 5 \cdot 3} \\
&= \frac{(-1)^n 2^n (2 \cdot 1)(2 \cdot 2) \cdots (2 \cdot n)}{(2n+1)!} \\
&= \frac{(-1)^n 2^{2n} n!}{(2n+1)!}.
\end{aligned}$$

Similarly, (10) determines a_n, for n even, in terms of a_0, which is arbitrary. We find that

$$a_{2n} = \frac{(-1)^n a_0}{n!}, \qquad (n \geq 1).$$

Hence y is an analytic solution of (4) if and only if y is given by

$$y = a_0 \sum_{n=0}^{\infty} \frac{(-1)^n}{n!} x^{2n} + \sum_{n=0}^{\infty} \frac{(-1)^n 2^{2n} n!}{(2n+1)!} x^{2n+1}. \qquad (11)$$

The ratio test shows that both series in (11) converge for all x. Therefore our thus far formal manipulations are justified, and there are analytic solutions of (4). The treatment of Section 2–4 would give us the solutions (11) in the form

$$y = a_0 e^{-x^2} + e^{-x^2} \int e^{x^2} dx.$$

EXAMPLE 2. Find the series in $(x - 1)$ for the solutions of

$$y' = y + x^2. \tag{12}$$

Suppose that (12) has an analytic solution

$$y = a_0 + a_1(x - 1) + a_2(x - 1)^2 + \cdots. \tag{13}$$

The function x^2 has the following series in powers of $(x - 1)$:

$$x^2 = 1 + 2(x - 1) + (x - 1)^2.$$

Hence (13) is a solution of (12) if we have the identity

$$a_1 + 2a_2(x - 1) + 3a_3(x - 1)^2 + \cdots$$
$$= a_0 + a_1(x - 1) + a_2(x - 1)^2 + \cdots + 1 + 2(x - 1) + (x - 1)^2$$
$$= (a_0 + 1) + (a_1 + 2)(x - 1)$$
$$\qquad + (a_2 + 1)(x - 1)^2 + a_3(x - 1)^3 + \cdots.$$

The coefficients must satisfy the relations

$$a_1 = a_0 + 1,$$
$$2a_2 = a_1 + 2, \qquad a_2 = \tfrac{1}{2}(a_0 + 3),$$
$$3a_3 = a_2 + 1, \qquad a_3 = \frac{1}{3!}(a_0 + 5),$$
$$4a_4 = a_3, \qquad a_4 = \frac{1}{4!}(a_0 + 5),$$
$$\vdots$$
$$na_n = a_{n-1}, \qquad a_n = \frac{1}{n!}(a_0 + 5), \quad (n \geq 3).$$

Conversely, if the series with the coefficients determined above converges, then it is a solution. The series is

$$y = a_0 + (a_0 + 1)(x - 1) + \tfrac{1}{2}(a_0 + 3)(x - 1)^2$$
$$+ \sum_{n=3}^{\infty} \frac{1}{n!}(a_0 + 5)(x - 1)^n, \tag{14}$$

which converges for all values of x.

The method of the preceding examples consists simply of substituting an arbitrary series in the differential equation and determining the coefficients. If only a few terms of the series are wanted, as an approximation to the solution, it is frequently easier to determine the coefficients directly from formula (11), Section 2–6.

EXAMPLE 3. Find the first five terms of the series in $(x - 1)$ for the solution y of

$$y' = x + y^2, \tag{15}$$

such that $y(1) = 1$.

Assuming that the solution of (15) such that $y(1) = 1$ is analytic around 1, the solution must be

$$y = \sum_{n=0}^{\infty} \frac{1}{n!} y^{(n)}(1)(x - 1)^n.$$

The initial condition gives the first coefficient, $a_0 = y(1) = 1$. The second coefficient, $a_1 = y'(1)$, can be determined from (15).

$$y'(1) = 1 + [y(1)]^2 = 1 + 1 = 2.$$

By differentiating (15), we get

$$
\begin{aligned}
y' &= x + y^2, & y'(1) &= 2, \\
y'' &= 1 + 2yy', & y''(1) &= 5, \\
y''' &= 2(y')^2 + 2yy'', & y'''(1) &= 18, \\
y^{(iv)} &= 4y'y'' + 2y'y'' + 2yy''', & y^{(iv)}(1) &= 96.
\end{aligned}
$$

Hence the first five terms of the series are

$$y = 1 + 2(x - 1) + \tfrac{5}{2}(x - 1)^2 + 3(x - 1)^3 + 4(x - 1)^4 + \cdots.$$

PROBLEMS

1. Find the series in x for the solutions of $y' = y$.
2. Find the series in x for the solution of $y' = y + x$, $y(0) = -1$.
3. Use (14) to show that the solution of (12) such that $y(1) = 0$ is $y = 5e^{(x-1)} - 5 - 4(x - 1) - (x - 1)^2$.
4. Find the series in x for the solutions of $y' + xy = x$. Show that the solutions are

$$y = 1 + (a_0 - 1) \sum_{n=0}^{\infty} \frac{1}{n!} \left(-\frac{x^2}{2}\right)^n = 1 + (a_0 - 1)e^{-x^2/2}.$$

5. Find the series in x for the solutions of $(1 - x)y' + y = 2x$. What is the interval of convergence of the series?

6. (a) Find all series in $(x - 2)$ for solutions of $(x - 2)y' + y = 2x - 1$.

(b) Find all solutions, and explain why only one series is found in part (a).

7. Find the series in x for the solutions of $y' + y = (x + 1)^2$.

8. Use (11), Section 2-6, to find the series in $(x - 1)$ for the solution of $y' = x^2 - y$ such that $y(1) = 0$. Show that the series equals $-e^{-(x-1)} + (x - 1)^2 + 1$.

9. Use (11), Section 2-6, to find the first five terms of the series in x for the solution of $y' = x + y^2$ such that $y(0) = 1$.

10. Use (11), Section 2-6, to find the first three nonzero terms of the solution of $y' = 1 + y^2$, $y(0) = 0$. Show that the solution is $y = \tan x$ and check the result by evaluating the first five derivatives of $\tan x$ at 0.

11. Find the series in x for the solutions of $y'' + y = 0$, and show [cf. (7) and (8) of Section 2-6] that they are $y = a_0 \cos x + a_1 \sin x$.

ANSWERS

1. $y = a_0 \sum_{n=0}^{\infty} \frac{1}{n!} x^n$

2. $y = -1 - x$

5. $y = a_0(1 - x) + \sum_{n=2}^{\infty} \frac{2}{n(n - 1)} x^n, \quad -1 < x < 1$

6. $y = (x^2 - x + c)/(x - 2)$. The only solution analytic around 2 is $y = 3 + (x - 2)$.

7. $y = a_0 + (1 - a_0)x + \frac{1}{2}(1 + a_0)x^2 + (1 - a_0) \sum_{n=3}^{\infty} (-1)^{n+1} \frac{1}{n!} x^n$

8. $y = (x - 1) + \frac{1}{2}(x - 1)^2 + \sum_{n=3}^{\infty} (-1)^{n+1} \frac{1}{n!} (x - 1)^n$

9. $y = 1 + x + \frac{3}{2!} x^2 + \frac{8}{3!} x^3 + \frac{34}{4!} x^4 + \cdots$

10. $y = x + \frac{2}{3!} x^3 + \frac{16}{5!} x^5 + \cdots$

11. $a_{2n} = (-1)^n \frac{1}{(2n)!} a_0, \quad a_{2n+1} = (-1)^n \frac{1}{(2n + 1)!} a_1$

CHAPTER 3

LINEAR EQUATIONS

3-1 Introduction. A *linear equation* is one of the form

$$y^{(n)} + p_{n-1}(x)y^{(n-1)} + \cdots + p_1(x)y' + p_0(x)y = q(x). \tag{1}$$

We will always assume that $p_0, p_1, \ldots, q$ are all defined and continuous on some interval (a, b). If the right-hand member, q, in (1) is zero, we say (1) is a *homogeneous* linear equation. The following are some examples of linear equations:

$$y' + xy = e^x,$$

$$y^{(iv)} + 3y' + e^x y = 0,$$

$$\theta'' + \frac{k}{m} \theta' + \frac{g}{l} \theta = 0,$$

$$s'' - \frac{k}{m} s' = g,$$

$$y'' = e^x,$$

$$y'' + e^x y' + x^2 y = \sin x.$$

The class of linear equations is large enough to encompass a great many of the most useful and frequently encountered differential equations. At the same time, the linear equation is sufficiently specialized to admit a very comprehensive and elegant theory. We will develop this theory in detail and see that it provides simple methods of solving a large class of of linear equations. Our principal tool will be the basic existence and uniqueness theorem for linear equations. As we indicated in Section 1-4, there is a general existence and uniqueness theorem for nth order equations of the form

$$y^{(n)} = F(x, y, y', \ldots, y^{(n-1)}). \tag{2}$$

However, Eq. (1) is much more restrictive than (2), and we should expect a better theorem for linear equations because of the stronger hypotheses inherent in the form of (1). Here is the theorem (to be proved in Chapter 7) for linear equations.

THEOREM 1. *If $p_0, p_1, \ldots, p_{n-1}, q$ are continuous on (a, b), and x_0 is any number in the interval (a, b), and $y_0, y_0', \ldots, y_0^{(n-1)}$ are any numbers,* *then there is a unique function y which is a solution of* (1) *on all of* (a, b) *such that*

$$y(x_0) = y_0, \qquad y'(x_0) = y_0', \ldots, y^{(n-1)}(x_0) = y_0^{(n-1)}.$$

The thing to note about this theorem is that it guarantees the existence of a solution on the whole interval on which the coefficient functions are continuous. Moreover, the uniqueness of the solution is an automatic consequence of the form of the equation (compare with the examples of Section 1–3).

We saw in Chapter 1 that the uniqueness aspect of such a theorem can be used to show that a given family of solutions contains *all* solutions. Theorem 1 says that there is exactly one solution y of (1) for any initial conditions

$$y(x_0) = y_0, \qquad y'(x_0) = y_0', \ldots, y^{(n-1)}(x_0) = y_0^{(n-1)}. \tag{3}$$

If we find a family of solutions of (1) such that *any* initial conditions (3) are satisfied by some member of the family, then the family contains all solutions. This idea is central in our development of the theory of linear equations.

Problems

1. The equation $y' = 1 + y^2$ is not linear, but the function $F(x, y) = 1 + y^2$ has all the continuity properties one could ask for. Show that nevertheless there is no function which is a solution on the interval $[0, 4]$. [*Hint:* Solve the equation.]

2. Suppose $q(x) \equiv 0$ in (1) and y is a solution of (1) on (a, b) such that for some x_0 in (a, b).

$$y(x_0) = y'(x_0) = \cdots = y^{(n-1)}(x_0) = 0.$$

Show that $y(x) \equiv 0$ on (a, b).

3. Write out Theorem 1 specifically for the case $n = 1$ and prove it (cf. Section 2–4).

4. Write out Theorem 1 specifically for the case $n = 2$.

3–2 Two theorems on linear algebraic equations. In trying to satisfy initial conditions for solutions of a linear differential equation we are forced to consider simultaneous linear algebraic equations. The theorems that are needed from algebra are given here.

Suppose a_{ij} for $i = 1, 2, \ldots, n$ and $j = 1, 2, \ldots, n$ are given numbers, and $b_1, \ldots, b_n$ are given numbers. We want to know when there are numbers $c_1, \ldots, c_n$ satisfying the equations

$$\begin{aligned}
c_1 a_{11} + c_2 a_{12} + \cdots + c_n a_{1n} &= b_1, \\
c_1 a_{21} + c_2 a_{22} + \cdots + c_n a_{2n} &= b_2, \\
&\vdots \\
c_1 a_{n1} + c_2 a_{n2} + \cdots + c_n a_{nn} &= b_n.
\end{aligned} \tag{1}$$

The number

$$D = \begin{vmatrix} a_{11} & a_{12} & \ldots & a_{1n} \\ a_{21} & a_{22} & \ldots & a_{2n} \\ \vdots & & & \\ a_{n1} & a_{n2} & \ldots & a_{nn} \end{vmatrix} \tag{2}$$

is called the *determinant of the system* (1).

THEOREM 1.   *If $D \neq 0$, there is a unique set of numbers $c_1, \ldots, c_n$ satisfying (1).*

If the numbers $b_1, \ldots, b_n$ in (1) are all zero, then (1) is called a system of *homogeneous* equations. Such a system always has at least the solution $c_1 = c_2 = \cdots = c_n = 0$. This is called the *trivial solution*, and we are generally interested in whether there are other (nontrivial) solutions.

THEOREM 2.   *If $b_1 = b_2 = \cdots = b_n = 0$, then (1) has a nontrivial solution (some $c_i \neq 0$) if and only if $D = 0$.*

PROBLEMS

1. Show that half of Theorem 2 is a consequence of Theorem 1.
2. Consider the family of functions $c_1 e^x + c_2 e^{-x}$. For what values of x is there a function y in this family such that $y(x) = 1$, $y'(x) = 2$?
3. Consider the family of functions $c_1 e^x + c_2 e^{-x} + c_3 \cosh x$. For what values of x is there a function y in this family such that $y(x) = 1$, $y'(x) = 0$, and $y''(x) = 0$?
4. Solve the system

$$3c_1 + c_2 + c_3 = 8,$$
$$c_1 - 2c_2 + c_3 = 0,$$
$$c_1 + c_2 - c_3 = 0.$$

5. Find a nontrivial solution of the system

$$c_1 + 2c_2 + c_3 = 0,$$
$$2c_1 - c_2 - c_3 = 0,$$
$$c_1 + 7c_2 + 4c_3 = 0.$$

ANSWERS

2. All values of x
3. No values of x, since $y''(x) \equiv y(x)$ for every function in the family.
4. $c_1 = 1$, $c_2 = 2$, $c_3 = 3$
5. $c_1 = 1$, $c_3 = -3$, $c_3 = 5$

3-3 General theory of linear equations.

We will consider the equation

$$y^{(n)} + p_{n-1}(x)y^{(n-1)} + \cdots + p_1(x)y' + p_0(x)y = q(x),$$

where $p_0, p_1, \ldots, p_{n-1}, q$ are defined and continuous on some interval (a, b). For simplicity, the independent variable x will be omitted from the coefficient functions $p_0, \ldots, p_{n-1}, q$ as well as from the dummy function y, and the equation will be written

$$y^{(n)} + p_{n-1}y^{(n-1)} + \cdots + p_1y' + p_0y = q. \tag{1}$$

The equation obtained from (1) by replacing q by zero,

$$y^{(n)} + p_{n-1}y^{(n-1)} + \cdots + p_1y' + p_0y = 0, \tag{2}$$

is called the *reduced equation* for (1), or the *homogeneous equation* associated with (1).

In this section we will not be concerned with specific solutions to specific examples of (1). Instead we investigate the general structure of the solutions of (1) and find that there is a simple and elegant theory inherent in the form of (1). We will show that to solve (1) it is necessary to find only one solution of (1) and all solutions of the reduced equation (2). Because of the linear form of (2), and the fact that the right-hand member is zero, any linear combination of solutions of this equation is again a solution. Using Theorem 1 of Section 3-1, we show that the n-parameter family which is the set of all linear combinations of any n "essentially different" solutions of (2) is the family of all solutions.

We proceed to fill in the details.

THEOREM 1. *If y_0 and y_1 are any solutions of* (1), *then* $y_1 = y_0 + u$, *where u is a solution of* (2).

Proof. Define $u = y_1 - y_0$. Then we must show that u is a solution of (2). Since $u' = y_1' - y_0'$, and $u'' = y_1'' - y_0''$, etc., we obtain, upon substituting u in the left side of (2),

$$(y_1^{(n)} - y_0^{(n)}) + p_{n-1}(y_1^{(n-1)} - y_0^{(n-1)}) + \cdots + p_0(y_1 - y_0)$$
$$= (y_1^{(n)} + p_{n-1}y_1^{(n-1)} + \cdots + p_0y_1)$$
$$\quad - (y_0^{(n)} + p_{n-1}y_0^{(n-1)} + \cdots + p_0y_0)$$
$$= q - q = 0.$$

THEOREM 2. *If y_0 is a solution of* (1) *and u is a solution of* (2), *then* $y_0 + u$ *is a solution of* (1).

Proof. Problem 1.

The following is a restatement of Theorems 1 and 2:

If y_0 is any solution of (1), *then the set of solutions of* (1) *consists of all functions of the form $y_0 + u$, where u is any solution of* (2). *That is, to find all solutions of* (1) *it is necessary and sufficient to find one solution y_0 of* (1) *and all solutions u of* (2).*

The following example illustrates the ideas which are developed in the sequel.

EXAMPLE 1. Consider the second order equation $y'' - y = 1 - x$. The function $y_0(x) = x - 1$ is a solution. The reduced equation is $y'' - y = 0$. By direct verification (Problem 2) one sees that $c_1 e^x + c_2 e^{-x}$ is a solution of the reduced equation for any numbers c_1 and c_2. To determine that the family $c_1 e^x + c_2 e^{-x}$ contains all solutions of the reduced equation, we check to see that any initial conditions $y(x_0) = y_0$, $y'(x_0) = y_0'$ are satisfied by some function in the family. For any numbers x_0, y_0, y_0' there must be numbers c_1 and c_2 such that

$$c_1 e^{x_0} + c_2 e^{-x_0} = y_0,$$
$$c_1 e^{x_0} - c_2 e^{-x_0} = y_0'.$$

The determinant of this system is

$$\begin{vmatrix} e^{x_0} & e^{-x_0} \\ e^{x_0} & -e^{-x_0} \end{vmatrix} = -2 \neq 0,$$

so there is a solution for c_1 and c_2. Theorem 1 says there is exactly one solution for any given initial conditions, and since there is one from the family $c_1 e^x + c_2 e^{-x}$, this family contains all solutions of the reduced equation. The solutions of $y'' - y = 1 - x$ are therefore the functions $x - 1 + c_1 e^x + c_2 e^{-x}$.

The following theorem allows us to look separately at the individual terms of the right-hand member q in finding a single solution of (1).

THEOREM 3 (Superposition principle). *If $q = q_1 + q_2$, and y_i is a solution of $y^{(n)} + p_{n-1} y^{(n-1)} + \cdots + p_0 y = q_i$, $(i = 1, 2)$, then $y_1 + y_2$ is a solution of* (1).

Proof. If $y_1 + y_2$ is substituted in the left side of (1), the terms containing y_1 and its derivatives can be separated from those containing y_2 and its derivatives. By hypothesis, the terms containing y_1 add up to q_1, and those containing y_2 add up to q_2.

Now we turn to the structure of the solutions of the reduced equation (2).

THEOREM 4. *If $y_1, \ldots, y_k$ are solutions of* (2), *then any linear combination $c_1 y_1 + \cdots + c_k y_k$ ($c_1, \ldots, c_k$ constants) is also a solution of* (2).

Proof. The theorem follows by induction once it is proved for any two functions and any two constants. So assume y_1 and y_2 are solutions of (2) and let $u = c_1 y_1 + c_2 y_2$. Then

$$u' = c_1 y_1' + c_2 y_2', \qquad u'' = c_1 y_1'' + c_2 y_2'', \quad \text{etc.}$$

Substituting u in (2), we get

$$
\begin{aligned}
u^{(n)} &+ p_{n-1} u^{(n-1)} + \cdots + p_0 u \\
&= (c_1 y_1^{(n)} + c_2 y_2^{(n)}) + p_{n-1}(c_1 y_1^{(n-1)} + c_2 y_2^{(n-1)}) + \cdots \\
&\qquad + p_0(c_1 y_1 + c_2 y_2) \\
&= c_1[y_1^{(n)} + p_{n-1} y_1^{(n-1)} + \cdots p_0 y_1] \\
&\qquad + c_2[y_2^{(n)} + p_{n-1} y_2^{(n-1)} + \cdots + p_0 y_2] \\
&= c_1 0 + c_2 0 = 0.
\end{aligned}
$$

EXAMPLE 2. The functions e^x, e^{-x}, $\sinh x$ are solutions of $y''' - y' = 0$. By Theorem 4, any linear combination $c_1 e^x + c_2 e^{-x} + c_3 \sinh x$ is also a solution. However, any one of the functions e^x, e^{-x}, $\sinh x$ can be written as a linear combination of the other two (e.g., $\sinh x = \frac{1}{2} e^x - \frac{1}{2} e^{-x}$), and the family of linear combinations is in reality a two-parameter family,

$$c_1 e^x + c_2 e^{-x} + c_3 \sinh x = C_1 e^x + C_2 e^{-x},$$

where $C_1 = c_1 + \frac{1}{2} c_3$ and $C_2 = c_2 - \frac{1}{2} c_3$. A set of initial conditions $y(x_0) = y_0$, $y'(x_0) = y_0'$, $y''(x_0) = y_0''$ for this family would consist of three linear equations in the two unknowns C_1 and C_2:

$$y(x_0) = C_1 e^{x_0} + C_2 e^{-x_0} = y_0,$$

$$y'(x_0) = C_1 e^{x_0} - C_2 e^{-x_0} = y_0',$$

$$y''(x_0) = C_1 e^{x_0} + C_2 e^{-x_0} = y_0''.$$

Such a system will not always have a solution, and the family does not contain all solutions.

Now consider in general the problem of satisfying any initial conditions with some function from a given family. Suppose $y_1, \ldots, y_n$ are solutions of the homogeneous equation (2), so that each function in the family

$$y = c_1 y_1 + \cdots + c_n y_n \tag{3}$$

is also a solution. Consider any set of initial conditions

$$y(x_0) = y_0, \qquad y'(x_0) = y_0', \ldots, y^{(n-1)}(x_0) = y_0^{(n-1)}. \tag{4}$$

If one of the functions (3) is to satisfy (4), we must have numbers $c_1, \ldots, c_n$ satisfying

$$\begin{aligned}
c_1 y_1(x_0) + \cdots + c_n y_n(x_0) &= y_0, \\
c_1 y_1'(x_0) + \cdots + c_n y_n'(x_0) &= y_0', \\
c_1 y_1^{(n-1)}(x_0) + \cdots + c_n y_n^{(n-1)}(x_0) &= y_0^{(n-1)}.
\end{aligned} \tag{5}$$

According to Theorem 1, Section 3–2, the system (5) is satisfied for some numbers $c_1, \ldots, c_n$ provided the determinant of the coefficients is not zero. For any set of $(n - 1)$ times differentiable functions $y_1, \ldots, y_n$, the determinant

$$W(y_1(x), \ldots, y_n(x)) = \begin{vmatrix} y_1(x) & y_2(x) & \ldots & y_n(x) \\ y_1'(x) & y_2'(x) & \ldots & y_n'(x) \\ y_1^{(n-1)}(x) & y_2^{(n-1)}(x) & \ldots & y_n^{(n-1)}(x) \end{vmatrix}$$

is called the *Wronskian* of the functions. Restating the remarks above, we see that the system (5) will have a solution for $c_1, \ldots, c_n$ provided $W(y_1(x_0), \ldots, y_n(x_0)) \neq 0$. If the family (3) is to satisfy *every* set of initial conditions (in particular, every x_0), we must have $W(y_1(x), \ldots, y_n(x)) \neq 0$ for all x. The situation is actually somewhat simpler than indicated, as the following definition and theorems show.

DEFINITION 1. Functions $u_1, \ldots, u_k$ are *linearly dependent* on an interval I if there are numbers $c_1, \ldots, c_k$, not all zero, such that $c_1 u_1(x) + \cdots + c_k u_k(x) \equiv 0$ on I. The functions $u_1, \ldots, u_k$ are *linearly independent* on I if they are not linearly dependent; i.e., if $c_1 u_1(x) + \cdots + c_k u_k(x) \equiv 0$ implies $c_1 = c_2 = \cdots = c_k = 0$.

EXAMPLE 3. *Linearly dependent functions.*

(A) e^x, e^{-x}, $\cosh x$; $\frac{1}{2}e^x + \frac{1}{2}e^{-x} - \cosh x \equiv 0$

(B) 0, x, e^x; $14 \cdot 0 + 0 \cdot x + 0 \cdot e^x \equiv 0$

(C) $x + 1$, $2x - 3$, 5; $2(x + 1) - (2x - 3) - 5 \equiv 0$

(D) 2, $\sin^2 x$, $\cos^2 x$; $1 \cdot 2 - 2 \cdot \sin^2 x - 2 \cdot \cos^2 x \equiv 0$

Linearly independent functions.

(E) 1, x, x^2, x^3 (G) e^x, xe^x, $\sin x$

(F) e^x, e^{2x}, e^{3x} (H) 1, $\cos x$

THEOREM 5. *If $u_1, \ldots, u_k$ are any $(k-1)$ times differentiable functions which are linearly dependent on I, then $W(u_1(x), \ldots, u_k(x)) \equiv 0$ on I.*

Proof. Let $c_1 u_1(x) + \cdots + c_k u_k(x) \equiv 0$ on (a, b) with not all $c_i = 0$. Differentiating $(k-1)$ times, we obtain the relations

$$
\begin{aligned}
c_1 u_1(x) + \cdots + c_k u_k(x) &\equiv 0, \\
c_1 u_1'(x) + \cdots + c_k u_k'(x) &\equiv 0, \\
c_1 u_1^{(k-1)}(x) + \cdots + c_k u_k^{(k-1)}(x) &\equiv 0.
\end{aligned} \tag{6}
$$

For any fixed x, this is a system of homogeneous algebraic equations in $c_1, \ldots, c_k$. By assumption, this system has a nontrivial solution for each x. By Theorem 2, Section 3–2, the determinant $W(u_1(x), \ldots, u_k(x)) = 0$ for all x.

COROLLARY. *If the Wronskian of any set of functions is nonzero for any x_0, then the functions are linearly independent on any interval containing x_0.*

The converse of Theorem 5 is false for arbitrary functions $u_1, \ldots, u_k$ (Problem 7). However, if we have n functions which are solutions of the nth order homogeneous equation (2), a statement even stronger than the converse of Theorem 5 is true.

THEOREM 6. *If $y_1, \ldots, y_n$ are n solutions of the nth order homogeneous equation (2), and $W(y_1(x_0), \ldots, y_n(x_0)) = 0$ for some x_0 in (a, b), then $y_1, \ldots, y_n$ are linearly dependent on (a, b), and hence $W(y_1(x), \ldots, y_n(x)) \equiv 0$ on (a, b).*

Proof. Recall (Problem 2, Section 3–1,) that the only solution of (2) satisfying $y(x_0) = y'(x_0) = \cdots = y^{(n-1)}(x_0) = 0$ is the function identically zero. Now suppose $W(y_1(x_0), \ldots, y_n(x_0)) = 0$. Then the system

$$
\begin{aligned}
c_1 y_1(x_0) + \cdots + c_n y_n(x_0) &= 0, \\
c_1 y_1'(x_0) + \cdots + c_n y_n'(x_0) &= 0, \\
&\vdots \\
c_1 y_1^{(n-1)}(x_0) + \cdots + c_n y_n^{(n-1)}(x_0) &= 0,
\end{aligned} \tag{7}
$$

has a nontrivial solution, $c_1, \ldots, c_n$ not all zero. For any $c_1, \ldots, c_n$ which are a nontrivial solution of (7), let $y = c_1 y_1 + \cdots + c_n y_n$. Then y is a solution of (2), and by (7), we have $y(x_0) = y'(x_0) = \cdots = y^{(n-1)}(x_0) = 0$. Therefore y is identically zero, which says that $y_1, \ldots, y_n$

are linearly dependent on (a, b). By Theorem 5 the Wronskian of $y_1, \ldots, y_n$ is identically zero.

COROLLARY. *The Wronskian of n solutions of (2) is identically zero, or is never zero.*

THEOREM 7. *The nth order equation (2) has n solutions which are linearly independent on (a, b). If $y_1, \ldots, y_n$ is any set of linearly independent solutions of (2), then the family*

$$c_1 y_1 + \cdots + c_n y_n \tag{8}$$

contains all solutions.

Proof. The fact that (8) contains all solutions is immediate from the Corollary to Theorem 6. The fact that the Wronskian is never zero for independent solutions allows us to satisfy any set of initial conditions as in (5). The uniqueness statement of Theorem 1, Section 3–1, shows that *the* solution for any set of initial conditions is in the family (8), and thus (8) contains all solutions. To see that there are n independent solutions, consider the n solutions $y_1, \ldots, y_n$ corresponding to the n sets of initial conditions $(i), \ldots, (n)$:

$(i)\quad y(x_0) = 1,\quad y'(x_0) = y''(x_0) = \cdots = y^{(n-1)}(x_0) = 0,$

$(ii)\quad y(x_0) = 0,\quad y'(x_0) = 1,\quad y''(x_0) = \cdots = y^{(n-1)}(x_0) = 0,$

$\vdots$

$(n)\quad y(x_0) = y'(x_0) = \cdots = y^{(n-2)}(x_0) = 0,\quad y^{(n-1)}(x_0) = 1.$

The Wronskian of $y_1, \ldots, y_n$ at x_0 is

$$\begin{vmatrix} 1 & 0 & 0 & \ldots & 0 \\ 0 & 1 & 0 & \ldots & 0 \\ \vdots & & & & \\ 0 & 0 & 0 & \ldots & 1 \end{vmatrix} = 1,$$

so these solutions are independent.

We will refer to (8) as the *general solution* of the homogeneous equation (2). Any specific function y_0 which is a solution of (1) will be called a *particular solution* of (1). The expression

$$y_0 + c_1 y + \cdots + c_n y_n, \tag{9}$$

which is the set of solutions of (1), will be called the *general solution* of (1).

PROBLEMS

1. Prove Theorem 2.

2. Verify (see Example 1) that $c_1e^x + c_2e^{-x}$ is a solution of $y'' - y = 0$ for all numbers c_1, c_2.

3. Show, as in Example 1, that every function in the family $c_1e^x + c_2xe^x$ is a solution of $y'' - 2y' + y = 0$, and that the family contains all solutions.

4. The functions $2 + x$, $x - e^x$, and $x + 1 + e^x$ are solutions of a certain nonhomogeneous second order linear equation.

   (a) Find (Theorem 1) two solutions of the reduced equation, neither of which is a constant multiple of the other.

   (b) Write a two-parameter family of solutions of the reduced equation (Theorem 4) and show that it contains all solutions of the reduced equation (cf. Example 1).

   (c) Find all solutions of the given nonhomogeneous equation (Theorems 1 and 2).

5. (See Example 2.) Find by inspection a third solution u of $y''' - y' = 0$ which is not a linear combination of e^x and e^{-x}. Show that the family $c_1u + c_2e^x + c_3e^{-x}$ contains all solutions.

6. (a) Show that $u_1, \ldots, u_k$ are linearly dependent if and only if some u_j can be written as a linear combination of the remaining functions.

   (b) Find three functions u_1, u_2, u_3 which are linearly dependent and such that u_1 cannot be expressed as a linear combination of u_2 and u_3.

   (c) Show that any set of functions containing the function identically zero is linearly dependent.

7. Show that x^3 and $|x|^3$ are not linearly dependent on $[-1, 1]$, but that $W(x^3, |x|^3) \equiv 0$. This shows that the converse of Theorem 5 is false.

8. Are x^3 and $|x|^3$ solutions on $[-1, 1]$ of (a) any second order homogeneous linear equation? (See problem 7.) (b) any third order homogeneous linear equation? (What is the third derivative of $|x|^3$ at 0?)

9. Which of these sets of functions are linearly dependent on the whole line?

   (a) $x - 3$, $6 - 2x$, e^x (b) $x + 1$, $2x - 3$, $3x + 4$
   (c) $\cos^2 x$, $\cos 2x$, $\sin^2 x$ (d) 1, e^x, xe^x

10. (a) Compute the Wronskians of the functions in (E), (F), (G), and (H) of Example 3.

   (b) Show (Theorem 6) that the functions in (H) are not solutions on $[-1, 1]$ of any second order homogeneous linear equation.

   (c) The functions of part (G) of Example 3 are solutions of $y'''' - 2y''' + 2y'' - 2y' + y = 0$. The Wronskian vanishes (at $\pi/2$) but is not identically zero. Why doesn't this contradict Theorem 6?

11. Suppose $y_1, \ldots, y_n$ are linearly independent solutions of (2), and y_0 is a solution of (1). Show directly that any initial conditions are satisfied by some function in the family $y_0 + c_1y_1 + \cdots + c_ny_n$.

12. Show that any $n + 1$ solutions of an nth order homogeneous linear equation are linearly dependent.

ANSWERS

4. (a) $2 + e^x$, $1 + 2e^x$; the condition of the problem rules out the choice of the zero function.

 (b) $c_1 + c_2e^x$; any initial conditions can be satisfied with a function from this family.

 (c) $x + c_1 + c_2e^x$

5. $u(x) = 1$

6. u_2 and u_3 must be linearly dependent

8. (a) No, (b) No

9. (a), (b), and (c)

10. (E) 12, (F) $2e^{6x}$, (G) $-2 \cos xe^{2x}$, (H) $-\sin x$

3–4 Second order equations with constant coefficients. One of the simplest and most useful cases of the linear equation is the second order equation

$$y'' + p_1y' + p_2y = q(x), \tag{1}$$

where p_1 and p_2 are constants. We will examine this equation in some detail and apply our results in Section 3–5 to some examples from mechanics, electricity, etc.

First, let us look at the reduced equation

$$y'' + p_1y' + p_2y = 0. \tag{2}$$

The problem is to find two linearly independent solutions. Since *two* functions are linearly dependent only if one is a constant multiple of the other (Problem 1), the check for independence can be made by inspection. Substituting $y = e^{rx}$ in the left side of (2), we get

$$e^{rx}(r^2 + p_1r + p_2). \tag{3}$$

Hence e^{rx} is a solution of (2) if r is a root of the algebraic equation

$$r^2 + p_1r + p_2 = 0. \tag{4}$$

The equation (4) is called the *auxiliary equation* for (1) or (2). If (4) has two real roots, r_1 and r_2, then e^{r_1x} and e^{r_2x} are the required two solutions of (2). If (4) has only one real root, r_0, then

$$r^2 + p_1r + p_2 = (r - r_0)^2 = r^2 - 2r_0r + r_0^2.$$

That is, $p_1 = -2r_0$ and $p_2 = r_0^2$. In this case the solutions of (2) are e^{r_0x} and xe^{r_0x} (Problem 2). If (4) has no real roots, the roots are conjugate complex numbers, say $a + ib$, $a - ib$, and

$$r^2 + p_1r + p_2 = [r - (a + ib)][r - (a - ib)]$$
$$= r^2 - 2ar + a^2 + b^2. \tag{5}$$

In other words,*

$$p_1 = -2a \quad \text{and} \quad p_2 = a^2 + b^2. \tag{6}$$

One can verify by substitution (Problem 3) that the solutions of (2) in this case are $e^{ax} \cos bx$ and $e^{ax} \sin bx$.

To summarize, the family of solutions of (2) is

$$c_1e^{r_1x} + c_2e^{r_2x}, \qquad \text{if } r_1 \text{ and } r_2 \text{ are distinct real roots of (4),}$$
$$c_1e^{r_0x} + c_2xe^{r_0x}, \qquad \text{if } r_0 \text{ is the only real root of (4),}$$
$$c_1e^{ax} \cos bx + c_2e^{ax} \sin bx, \qquad \text{if } a \pm ib \text{ are the roots of (4).}$$

EXAMPLE 1. (A) $y'' + y' - 2y = 0$.

The auxiliary equation is $r^2 + r - 2 = 0$, or $(r + 2)(r - 1) = 0$. The roots are 1 and -2, so the solutions of (A) are $c_1e^x + c_2e^{-2x}$.

(B) $y'' + 4y' + 4y = 0$.

The auxiliary equation is $r^2 + 4r + 4 = 0$, or $(r + 2)^2 = 0$. The single root is -2, so the solutions of (B) are $y = c_1e^{-2x} + c_2xe^{-2x}$.

(C) $y'' - 2y' + 5y = 0$.

The auxiliary equation is $r^2 - 2r + 5 = 0$. Since the discriminant is $(-2)^2 - 4(1)(5) = -16 < 0$, the roots are complex numbers $a \pm ib$. Comparing with (6), we see that $a = -\frac{1}{2}(-2) = 1$, and $b = \sqrt{5 - 1^2} = 2$. The solutions of (C) are $y = c_1e^x \cos 2x + c_2e^x \sin 2x$.

Having found all the solutions of (2), we are still faced with the problem of finding one solution of (1). Later we will give the so-called variation of parameters method for finding a solution of any linear equation when all the solutions of the reduced equation are known. For now we consider the simpler method of undetermined coefficients, which works whenever the right member q of (1) has the form

$$P(x)e^{ax} \cos bx + Q(x)e^{ax} \sin bx, \tag{7}$$

---

* Note that formulas (6) give an easy way to find a and b when (4) has the complex roots $a \pm ib$; $a = -\frac{1}{2}p_1$, and $b = \sqrt{p_2 - a^2}$.

with P and Q polynomials. Some examples of functions of the form (7) are:

$$x^3 - 2x + 1 \qquad (a = b = 0, P(x) = x^3 - 2x + 1),$$
$$(2 - x)e^x \qquad (a = 1, b = 0, P(x) = 2 - x),$$
$$3\cos 2x \qquad (a = 0, b = 2, P(x) = 3, Q(x) = 0),$$
$$x^2 \sin x \qquad (a = 0, b = 1, P(x) = 0, Q(x) = x^2),$$
$$3\cos x - x\sin x \qquad (a = 0, b = 1, P(x) = 3, Q(x) = x),$$
$$2xe^x \cos 2x \qquad (a = 1, b = 2, P(x) = 2x, Q(x) = 0).$$

The facts behind the method of undetermined coefficients are stated in the following theorem, which is proved later as a simple consequence of the theory of operators.

THEOREM 1 (Method of undetermined coefficients). *Let* P_n, Q_n, P_n^*, Q_n^* *be polynomials of degree* n *or less.*

If $q(x) = P_n(x)e^{ax}$, *then* (1) *has a solution of the form* $y = x^k P_n^*(x)e^{ax}$, *where* k *is* 0, 1, *or* 2; $k = 0$ *if* e^{ax} *is not a solution of the reduced equation* (2); $k = 1$ *if* e^{ax} *is a solution of* (2) *and* xe^{ax} *is not;* $k = 2$ *if both* e^{ax} *and* xe^{ax} *are solutions of* (2).

If $q(x) = P_n(x)e^{ax}\cos bx$, *or* $q(x) = Q_n(x)e^{ax}\sin bx$, *or* $q(x) = P_n(x)e^{ax}\cos bx + Q_n(x)e^{ax}\sin bx$, *with* $b \neq 0$, *then* (1) *has a solution of the form* $y = x^k(P_n^*(x)e^{ax}\cos bx + Q_n^*(x)e^{ax}\sin bx)$ *where* k *is* 0 *or* 1; $k = 0$ *if* $e^{ax}\cos bx$ *is not a solution of* (2), *and* $k = 1$ *if* $e^{ax}\cos bx$ *is a solution of* (2).

The "method" consists simply of substituting in the equation a function of the appropriate form with arbitrary polynomials $P_n^*(x) = A_0 + A_1 x + \cdots + A_n x^n$, $Q_n^*(x) = B_0 + B_1 x + \cdots + B_n x^n$, and seeing what the coefficients A_0, A_1, ..., B_0, B_1, ... must be for the function to be a solution. Note that if q contains either a *sine* or a *cosine* term, the trial function must contain both, and the polynomial coefficients P_n^*, Q_n^* must both be of the degree which is the maximum of the degrees of P_n and Q_n. The conditions on the factor x^k can be remembered as follows. A trial solution of the same form as q is multiplied by x or x^2, if necessary, so that none of the terms of the resulting function are solutions of the reduced equation. For example, if $q(x) = xe^x$, the trial solution of the same form is $(A + Bx)e^x$. If e^x is a solution of the reduced equation, there is no point in including the term Ae^x in the trial function, and the appropriate function is $(Ax + Bx^2)e^x$. If both e^x and xe^x are solutions of the reduced equation, the trial function should be

$$x^2(A + Bx)e^x = (Ax^2 + Bx^3)e^x.$$

EXAMPLE 2.   (i) $y'' - y' - 2y = 1 - x^2$,

(ii) $y'' - y' = 1 - x^2$.

The solutions of the reduced form of (i) are $c_1e^{-x} + c_2e^{2x}$. There is a solution of (i) of the form

$$y = (A + Bx + Cx^2)e^{0x} = A + Bx + Cx^2.$$

Substitution gives

$$2C - (B + 2Cx) - 2(A + Bx + Cx^2)$$

or

$$(2C - B - 2A) + (-2B - 2C)x - 2Cx^2.$$

We must have $-2C = -1$, $-2B - 2C = 0$, and $2C - B - 2A = 1$. This gives $C = \frac{1}{2}$, $B = -\frac{1}{2}$, and $A = \frac{1}{4}$. The solutions of (i) are

$$y = \frac{1}{4} - \frac{1}{2}x + \frac{1}{2}x^2 + c_1e^{-x} + c_2e^{2x}.$$

The solutions of the reduced form of (ii) are $c_1 + c_2e^x$. Since constants are solutions of the reduced equation, the trial solution for (ii) is $y = Ax + Bx^2 + Cx^3$. The coefficients A, B, C are evaluated as above.

EXAMPLE 3.   (i) $y'' - y' - 2y = e^x$,

(ii) $y'' - y' - 2y = (1 + x^2)e^x$,

(iii) $y'' - 3y' + 2y = 2e^x$,

(iv) $y'' - 2y' + y = (2 + x)e^x$.

In (i) and (ii), e^x is not a solution of the reduced equation, and the trial solutions are respectively $y = Ae^x$, and $y = (A + Bx + Cx^2)e^x$. In (iii) e^x is a solution of the reduced equation, and the trial solution is $y = Axe^x$. In (iv), e^x and xe^x are solutions of the reduced equation, so the trial solution is $y = x^2(A + Bx)e^x$. For example, in (iii), substitution of $y = Axe^x$ gives $A(2 + x)e^x - 3A(1 + x)e^x + 2Axe^x = -Ae^x$. Therefore Axe^x is a solution if $A = -2$, and the set of all solutions is $(c_1 - 2x)e^x + c_2e^{2x}$.

EXAMPLE 4.   (i) $y'' - y' - 2y = \cos x$,

(ii) $y'' + y = \cos x$,

(iii) $y'' + y = x \cos x$.

Here $\cos x$ and $\sin x$ are not solutions of the reduced equation in (i), but are in (ii) and (iii). The trial solutions for the three equations are

(i) $y = A \cos x + B \sin x$,

(ii) $y = Ax \cos x + Bx \sin x$,

(iii) $y = (Ax + Bx^2) \cos x + (Cx + Dx^2) \sin x$.

EXAMPLE 5. (i) $y'' + y = e^x \sin 2x$,

(ii) $y'' - 2y' + 5y = 3xe^x \cos 2x$,

(iii) $y'' - 2y' + 5y = xe^x \cos 2x - e^x \sin 2x$.

Here $e^x \cos 2x$ and $e^x \sin 2x$ are not solutions of the reduced equation for (i), but are for (ii) and (iii). The trial solutions are

(i) $y = Ae^x \cos 2x + Be^x \sin 2x$,

(ii) $y = (Ax + Bx^2)e^x \cos 2x + (Cx + Dx^2)e^x \sin 2x$,

(iii) same as (ii).

The superposition principle (Theorem 3, Section 3–3) can be used to extend the method of undetermined coefficients to any equation in which the right member is a sum of functions of the form (7).

EXAMPLE 6. $y'' - y' - 2y = 1 - x^2 + 2e^x - \cos x$.

We consider separately the equations $y'' - y' - 2y = 1 - x^2$ [Example 2(i)], $y'' - y' - 2y = 2e^x$ [Example 3(i)], and $y'' - y' - 2y = -\cos x$ [Example 4(i)]. Solutions of these three equations are respectively $\frac{1}{4} - \frac{1}{2}x + \frac{1}{2}x^2$, $-e^x$, and $\frac{3}{10} \cos x + \frac{1}{10} \sin x$. The solutions of the given equation are therefore

$$y = c_1 e^{-x} + c_2 e^{2x} + \tfrac{1}{4} - \tfrac{1}{2}x + \tfrac{1}{2}x^2 - e^x + \tfrac{3}{10} \cos x + \tfrac{1}{10} \sin x.$$

PROBLEMS

1. Show that *two* functions are linearly dependent if and only if one is a constant multiple of the other.

2. Show that if (4) has only one real root r_0, then $e^{r_0 x}$ and $xe^{r_0 x}$ are solutions of (2).

3. Show that if (4) has the roots $a + ib$ and $a - ib$, so that (2) has the form $y'' - 2ay' + (a^2 + b^2)y = 0$, then $e^{ax} \cos bx$ and $e^{ax} \sin bx$ are solutions of (2).

4. Solve the following equations.

(a) $y'' + y' + y = 0$

(b) $y'' - 3y' + 2y = 0$

(c) $y'' + y' - 6y = 0$

Find all solutions for the following and specify which solution satisfies the given initial conditions.

5. (a) $y'' = 0$, $y(2) = 2$, $y'(2) = 3$

(b) $y'' - 2y' + y = 0$, $y(0) = 1$, $y'(0) = 2$

(c) $y'' + 4y = 0$, $y(0) = 1$, $y'(0) = -2$

(d) $y'' - 2y' = 0$, $y(0) = 1$, $y'(0) = 4$

6. $y'' + 3y' = 6x$, $y(0) = 1$, $y'(0) = \frac{7}{3}$

7. $y'' + y = e^x$, $y(0) = 1$, $y'(0) = 0$

8. $y'' - 2y' + y = -25 \sin 2x$, $y(0) = -4$, $y'(0) = 8$

9. $y'' - 2y' + y = e^x$, $y(1) = \frac{3}{2}e$, $y'(1) = \frac{5}{2}e$

Find all solutions for Problems 10 through 14.

10. $y'' - 3y' + 2y = x^2 + 1$ 11. $y'' + y' + y = x + 2 + 3e^x$

12. $y'' + y' - 2y = e^x$ 13. $y'' - 2y' + 2y = e^x \cos x$

14. $y'' - 2y' = 4x + e^{3x}$

Write the form of a single solution for each of the following, but leave the coefficients undetermined.

15. $y'' - 4y = (x^3 - 2x)e^{2x} + x \cos x$

16. $y'' + 2y' - 3y = xe^x + e^{-3x} \sin x + e^x$

17. $y'' + 4y' + 8y = xe^{-2x}(3 + \sin 2x)$

18. (a) Prove that if $q(x) = a_0 + a_1x + \cdots + a_nx^n$, and $p_2 \neq 0$, then (1) has a solution $y = A_0 + A_1x + \cdots + A_nx^n$. [*Hint:* Start with the coefficient of x^n after substitution and work backwards.]

    (b) Consider the cases $p_2 = 0$, $p_1 \neq 0$, and $p_1 = p_2 = 0$.

    (c) Show that if $q(x) = (a_0 + a_1x + \cdots + a_nx^n)e^{ax}$, the substitution $y(x) = e^{ax}z(x)$ changes (1) to a linear equation in z with a polynomial right member. Use this and parts (a) and (b) to prove the first half of Theorem 1.

ANSWERS

4. (a) $y = c_1e^{-(1/2)x} \cos \dfrac{\sqrt{3}}{2} x + c_2e^{-(1/2)x} \sin \dfrac{\sqrt{3}}{2} x$

    (b) $y = c_1e^x + c_2e^{2x}$

    (c) $y = c_1e^{2x} + c_2e^{-3x}$

5. (a) $y = c_1 + c_2x$, $y = -4 + 3x$

    (b) $y = (c_1 + c_2x)e^x$, $y = (1 + x)e^x$

    (c) $y = c_1 \cos 2x + c_2 \sin 2x$, $y = \cos 2x - \sin 2x$

    (d) $y = c_1 + c_2e^{2x}$, $y = -1 + 2e^{2x}$

6. $y = c_1 + c_2e^{-3x} - \frac{2}{3}x + x^2$, $y = 2 - e^{-3x} - \frac{2}{3}x + x^2$

7. $y = c_1 \cos x + c_2 \sin x + \frac{1}{2}e^x$, $y = \frac{1}{2}(e^x + \cos x - \sin x)$

8. $y = (c_1 + c_2x)e^x - 4 \cos 2x + 3 \sin 2x$,
    $y = -4 \cos 2x + 3 \sin 2x + 2xe^x$

9. $y = (c_1 + c_2x + \frac{1}{2}x^2)e^x$, $y = (1 + \frac{1}{2}x^2)e^x$

10. $y = \frac{9}{4} + \frac{3}{2}x + \frac{1}{2}x^2 + c_1e^x + c_2e^{2x}$

11. $y = 1 + x + e^x + e^{-(1/2)x}\left(c_1 \cos \dfrac{\sqrt{3}}{2} x + c_2 \sin \dfrac{\sqrt{3}}{2} x\right)$

12. $y = (c_1 + \frac{1}{3}x)e^x + c_2e^{-2x}$

13. $y = \frac{1}{2}xe^x \sin x + e^x(c_1 \cos x + c_2 \sin x)$

14. $y = -x - x^2 + \frac{1}{3}e^{3x} + c_1 + c_2e^{2x}$

15. $y = x(A + Bx + Cx^2 + Dx^3)e^{2x} + (E + Fx) \cos x + (G + Hx) \sin x$

16. $y = x(A + Bx)e^x + Ce^{-3x} \sin x + De^{-3x} \cos x$

17. $y = x(A + Bx)e^{-2x} \sin 2x + x(C + Dx)e^{-2x} \cos 2x + (E + Fx)e^{-2x}$

3–5 Applications. Many of the common differential equations which arise in applications are of the type

$$y'' + p_1y' + p_2y = q(x) \qquad (p_1, p_2 \text{ constants}), \tag{1}$$

which was treated in the preceding section. In this section we will examine the differential equations of some simple physical systems and try to interpret the solutions in physical terms.

As a first example, consider a block of mass m attached to a spring and sliding on a horizontal table (Fig. 3–1). Let the spring constant be b^2 lb/ft so that the spring exerts a force of $-b^2s$ lb when the spring is stretched ($s > 0$) or compressed ($s < 0$) s ft. Assume that the assorted frictional effects exert a force proportional to the speed $s' = ds/dt$, and of course opposing the motion. This so-called damping force can be written $-2as'$, with $a \geq 0$. Since mass times acceleration ($s'' = d^2s/dt^2$) equals the total force, we have the equation $ms'' = -2as' - b^2s$, or

$$s'' + \frac{2a}{m} s' + \frac{b^2}{m} s = 0. \tag{2}$$

If in addition some external force $F(t)$ acts on the mass, the equation becomes

$$s'' + \frac{2a}{m} s' + \frac{b^2}{m} s = \frac{1}{m} F(t). \tag{3}$$

As an example of (3), let the mass hang from the spring (Fig. 3–2), so there is a constant force $F(t) = mg$ exerted by gravity.

The equation (3) also appears in the study of electric circuits. The current i in a series circuit containing a resistance R, inductance L, and

Motion ⟶

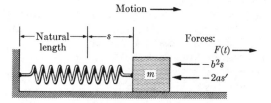

FIGURE 3–1

capacitance C is determined by the equation

$$\frac{d^2i}{dt^2} + \frac{R}{L}\frac{di}{dt} + \frac{1}{LC}i = \frac{1}{L}E'(t), \tag{4}$$

where $E(t)$ is the electromotive force applied at time t. If we have the equalities

$$R = 2a, \quad L = m, \quad C = \frac{1}{b^2}, \quad E' = F, \tag{5}$$

then Eqs. (3) and (4) are identical except for the dummy function used. This formal similarity between electric circuits and vibrating mechanical systems makes it possible to study the behavior of a mechanical system by setting up the corresponding circuit [by (5)] and measuring the current directly.

Now let us examine the solutions of (2), which will be added to any particular solution of (3), or with the changes (5), to any solution of (4). Physically, the solutions of (2) represent the motion which results if the system of Fig. 3-1 is set into motion and then released with no further force applied. These solutions are called *transients*, since the motion they represent dies out as t increases. The roots of the auxiliary equation for (2) are

$$\frac{-a + \sqrt{a^2 - b^2m}}{m} \quad \text{and} \quad \frac{-a - \sqrt{a^2 - b^2m}}{m}. \tag{6}$$

The nature of the solutions will depend on the sign of $a^2 - b^2m$; that is, on the relative magnitudes of the damping force and the restoring force times the mass. Let us consider separately the cases $a^2 > b^2m$, $a^2 = b^2m$, and $a^2 < b^2m$.

Case I $(a^2 > b^2m$; overdamping). The roots (6) are distinct real numbers, and the solutions of (2) are the functions

$$s = c_1 \exp\left(\frac{-a + \sqrt{a^2 - b^2m}}{m}\right)t + c_2 \exp\left(\frac{-a - \sqrt{a^2 - b^2m}}{m}\right)t. \tag{7}$$

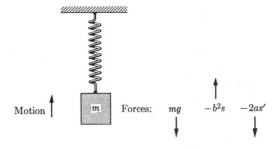

Motion | Forces: mg $-b^2s$ $-2as'$

FIGURE 3-2

Since $a > \sqrt{a^2 - b^2m}$, both terms in (7) are of the form ce^{-pt}, $p > 0$, and tend monotonically to zero as $t \to \infty$. The functions (7) change sign once or not at all (Problem 1), depending on the initial conditions.

Case II ($a^2 = b^2m$; critical damping). The auxiliary equation has one real root, $-a/m$, and the solutions of (2) are

$$s = (c_1 + c_2t)e^{-(a/m)t}. \tag{8}$$

The exponential is always positive, so each function (8) changes sign at most once. In both Cases I and II the damping force predominates, and there is no oscillation. The mass of Fig. 3–1 crosses over the equilibrium point at most once, then returns toward the equilibrium position. This kind of behavior is illustrated by the wheels of an automobile acting under the influence of the springs and shock absorbers.

Case III ($a^2 < b^2m$; underdamping). The roots (6) of the auxiliary equation are the complex numbers $-a/m \pm i\omega$, where

$$\omega = \frac{1}{m} \sqrt{mb^2 - a^2}. \tag{9}$$

The solutions of (2) are

$$s = e^{-(a/m)t}[c_1 \cos \omega t + c_2 \sin \omega t]. \tag{10}$$

If we define new constants A and α by

$$A = \sqrt{c_1^2 + c_2^2}, \quad \sin \alpha = \frac{c_1}{\sqrt{c_1^2 + c_2^2}}, \quad \cos \alpha = \frac{c_2}{\sqrt{c_1^2 + c_2^2}}, \tag{11}$$

then (10) can be written

$$s = \sqrt{c_1^2 + c_2^2} \, e^{-(a/m)t} \left\{ \frac{c_1}{\sqrt{c_1^2 + c_2^2}} \cos \omega t + \frac{c_2}{\sqrt{c_1^2 + c_2^2}} \sin \omega t \right\}$$

$$= Ae^{-(a/m)t} \sin (\omega t + \alpha). \tag{12}$$

The solutions are damped sine curves (Fig. 3–3). The motion is an oscillation about the equilibrium point with the amplitude of the vibrations decreasing to zero as t increases. The *period* of the vibration, or time required for a complete cycle, is $2\pi/\omega$ sec. The *frequency* is $\omega/2\pi$ cps.

In case the damping effects are negligible, $a = 0$, we have what is called *simple harmonic motion*. The equation (2) in this case is

$$s'' + \frac{b^2}{m} s = 0, \tag{13}$$

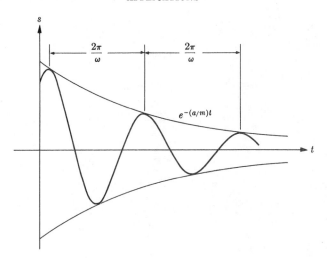

FIGURE 3–3

and the solutions are pure sine curves

$$s = c_1 \cos \frac{bt}{\sqrt{m}} + c_2 \sin \frac{bt}{\sqrt{m}},$$

or

$$s = A \sin \left(\frac{bt}{\sqrt{m}} + \alpha \right). \tag{14}$$

Now let us consider the situation in which the vibrations are forced, as in (3) or (4). The solutions we have found for (2) describe transient effects; the motion they represent is superimposed on the motion (a particular solution) which corresponds to the forcing function. The type of forcing function which can realistically be considered in (3) or (4) is of course limited; an unbounded function, for example, would not be reasonable. We will consider a forcing function of the form $F(t) = E \sin(\omega_0 t)$ which is the case, for instance, when an alternating electromotive force is applied to the circuit (4). The equation we consider is therefore

$$s'' + \frac{2a}{m} s' + \frac{b^2}{m} s = \frac{E}{m} \sin(\omega_0 t). \tag{15}$$

By Theorem 1, Section 3–4, there is a solution of (15) of the form

$$s = A_1 \cos(\omega_0 t) + A_2 \sin(\omega_0 t),$$

or equivalently, of the form

$$s = A \sin (\omega_0 t - \alpha). \tag{16}$$

Substitution of (16) in the left side of (15) gives

$$A \left\{ \frac{b^2 - m\omega_0^2}{m} \sin (\omega_0 t - \alpha) + \frac{2a\omega_0}{m} \cos (\omega_0 t - \alpha) \right\}. \tag{17}$$

Now let

$$K = \sqrt{\left(\frac{b^2 - m\omega_0^2}{m} \right)^2 + \left(\frac{2a\omega_0}{m} \right)^2}$$

$$= \frac{1}{m} \sqrt{(b^2 - m\omega_0^2)^2 + 4a^2\omega_0^2},$$

and write (17) in the form

$$AK \left\{ \frac{b^2 - m\omega_0^2}{mK} \sin (\omega_0 t - \alpha) + \frac{2a\omega_0}{mK} \cos (\omega_0 t - \alpha) \right\}. \tag{18}$$

The quantity K is chosen so that

$$\left(\frac{b^2 - m\omega_0^2}{mK} \right)^2 + \left(\frac{2a\omega_0}{mK} \right)^2 = 1,$$

and hence there is a number α such that

$$\cos \alpha = \frac{b^2 - m\omega_0^2}{mK} \quad \text{and} \quad \sin \alpha = \frac{2a\omega_0}{mK}. \tag{19}$$

For this value of α, (18) becomes

$$AK \sin (\omega_0 t). \tag{20}$$

Therefore (16) is a solution of (15) provided α is determined by (19) and $AK = E/m$; that is, if

$$A = \frac{E}{\sqrt{(b^2 - m\omega_0^2)^2 + 4a^2\omega_0^2}}. \tag{21}$$

For a forcing function $F(t) = E \sin (\omega_0 t)$, the solutions will all approximate, as the transients die out, a sine curve with the same frequency $\omega_0/2\pi$ as the forcing function. The maximum displacement, as one would expect, will not occur at the same time as the maximum force, but will lag by an interval of α/ω_0 seconds.

If there is little damping (a is small) and the frequency $\omega_0/2\pi$ of the driving force approaches the natural damped frequency $\omega/2\pi$ [Formula (9)] of the system, then the amplitude A can be very large. (See Problem 3.) This is the phenomenon called *resonance*.

EXAMPLE 1. (See Fig. 3–2.) If a block weighing 32 lb is suspended from a spring, the spring stretches 16 in. ($\frac{4}{3}$ ft). Suppose the damping force in pounds equals twice the numerical value of the speed in feet per second. Find the position of the block at time t if the block is released at $t = 0$ with zero velocity and with the spring at its natural length.

The equation is (3) with $F(t) = 32$, $2a = 2$, and $m = 1$ (taking $g = 32$ ft/sec^2). The spring constant b^2 is given by $\frac{4}{3}b^2 = 32$. Hence $b^2 = 24$, and the equation is

$$s'' + 2s' + 24s = 32,$$

with the initial conditions

$$s(0) = 0, \qquad s'(0) = 0.$$

The roots of the auxiliary equation are

$$-1 + \sqrt{23}\, i \qquad \text{and} \qquad -1 - \sqrt{23}\, i,$$

and the solutions of the reduced equation are

$$s = c_1 e^{-t} \cos (\sqrt{23}\, t) + c_2 e^{-t} \sin (\sqrt{23}\, t).$$

A particular solution of the equation is clearly the constant $s = \frac{4}{3}$. When $t = 0$, $s = \frac{4}{3} + c_1 = 0$, so $c_1 = -\frac{4}{3}$. The velocity at time t is given by

$$s' = -\tfrac{4}{3}[-e^{-t} \cos (\sqrt{23}\, t) - \sqrt{23}\, e^{-t} \sin (\sqrt{23}\, t)]$$
$$+ c_2[-e^{-t} \sin (\sqrt{23}\, t) + \sqrt{23}\, e^{-t} \cos (\sqrt{23}\, t)].$$

When $t = 0$, $s' = -\frac{4}{3}(-1) + c_2\sqrt{23} = 0$, so $c_2 = -4/(3\sqrt{23})$. The position s at time t is

$$s = \tfrac{4}{3} - \tfrac{4}{3}e^{-t} \left[\cos (\sqrt{23}\, t) + \frac{1}{\sqrt{23}} \sin (\sqrt{23}\, t) \right].$$

EXAMPLE 2. The forces on a pendulum of mass m and length L are shown in Fig. 3–4. The force mg of gravity can be resolved into a component along the length of the pendulum, and a component $mg \sin \theta$ in the direction of the motion. We measure distance s along the arc of the

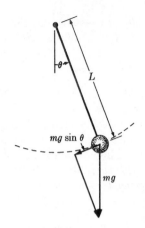

FIGURE 3–4

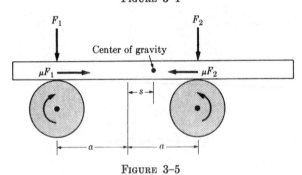

FIGURE 3–5

pendulum, so $s = L\theta$, and $s'' = L\theta''$. If we assume that there is no friction, the equation is

$$ms'' = -mg \sin \theta, \quad \text{or} \quad \theta'' + \frac{g}{L} \sin \theta = 0.$$

For small values of θ, $\sin \theta$ can usefully be approximated by θ, and we can assume the pendulum satisfies the equation

$$\theta'' + \frac{g}{L} \theta = 0.$$

This is the simple harmonic motion of (13), and the motion is periodic with constant frequency $(2\pi\sqrt{L/g})^{-1}$. To study motion other than small oscillations about $\theta = 0$, the exact equation must be used.

EXAMPLE 3. A bar of weight W lies across two counter-rotating cylinders as in Figure 3–5. If the coefficient of friction between the bar and the rollers is μ, then the left cylinder exerts a horizontal force of μF_1, and the right cylinder exerts a horizontal force of $-\mu F_2$, where F_1 and F_2 are the vertical forces on the rollers due to the weight of the bar. Since $F_1 + F_2 = W$, and $(a + s)F_1 = (a - s)F_2$, the equation describing the position of the bar's center of gravity is (Problem 5)

$$s'' + \frac{g\mu}{a}\, s = 0,$$

and the motion is simple harmonic.

Problems

1. Show that the functions (7) change sign once for positive t if $c_1/c_2 < -1$ and do not change sign if $c_1/c_2 \geq -1$. [Hint: consider the function $s = c_1 e^{-pt} + c_2 e^{-qt}$, where $0 < p < q$, and write $s = c_1 e^{-pt}[1 + (c_2/c_1)e^{(p-q)t}]$.]

2. Suppose the mass of Fig. 3–1 satisfies equation (2) and $a^2 = b^2 m$. Express s in terms of the initial position s_0 and initial velocity s_0'. Suppose $s_0 > 0$: For what initial velocities s_0' will the mass fail to cross the equilibrium point?

3. Find the amplitude A [formula (11)] of the solution of (15) when the frequency ω_0 of the driving force equals the natural damped frequency [formula (9)] of the system.

4. Write the solution of the equation of Example 1 in the form $s = \frac{4}{3} + A e^{-t}\sin(\sqrt{23}\,t + \alpha)$; that is, find A and α so this formula satisfies the initial conditions $s(0) = 0 = s'(0)$.

5. Complete the derivation of the equation of Example 3. What is the maximum velocity the center of gravity of the bar can have midway between the rollers without the bar falling off?

6. What length pendulum (Example 2) will swing from one side to the other each second? (Take $g = 32$ ft/sec^2.)

7. A cylinder of mass m has a cross section area of 2 ft^2. The cylinder floats with its axis vertical in water of density ρ lb/ft. Show that if the cylinder is disturbed from equilibrium, it bobs up and down in simple harmonic motion. Find the frequency of the motion. (The bouyant force is equal to the weight of water displaced.)

8. A 16-ft chain weighing 2 lb/ft (total mass equals 1) rests on a table with part of the chain hanging over the edge of the table. The coefficient of friction between the chain and table is $\frac{1}{3}$, so that friction exerts a retarding force of $\frac{1}{3} \cdot (16 - x) \cdot 2$ lb when x feet of chain are over the edge of the table. Suppose the chain is released with 6 ft hanging over the edge. How long does it take the chain to slide off?

9. A block weighing 64 lb hangs from a spring as in Fig. 3–2, and stretches the spring 6 in. The block is also attached to a shock absorber which exerts a force of $4v$ lb, where v is the speed in feet per second. If the block receives an impact giving it an initial velocity of 20 ft/sec downward at the equilibrium position, find the position at time t. (Measure s downward from the natural length of the spring, and use $g = 32$ ft/sec^2.)

ANSWERS

2. $s = \left[s_0 + \left(s_0' + \dfrac{as_0}{m} \right) t \right] e^{-(a/m)t}; \quad s_0' \geq - \dfrac{as_0}{m}$

3. $A = \dfrac{Em}{a\sqrt{a^2 + 4m^2\omega_0^2}}$

4. $A = -\dfrac{8}{3} \sqrt{\dfrac{6}{23}} \doteq -1.36, \quad \alpha \doteq 1.37$

5. $s_0' = \sqrt{ag\mu}$ 6. 3.24 ft

7. $\dfrac{1}{\pi} \sqrt{\dfrac{\rho}{2m}}$ cps   8. $\sqrt{\dfrac{3}{8}} \cosh^{-1} 6$ sec

9. $s = \dfrac{1}{2} + \dfrac{20}{\sqrt{63}} e^{-t} \sin (\sqrt{63}\, t)$

CHAPTER 4

SPECIAL METHODS FOR LINEAR EQUATIONS

4–1 Complex-valued solutions. The treatment of homogeneous linear differential equations with constant coefficients is much simplified if we consider complex solutions. We first review some of the basic facts about complex functions.

A complex-valued function f of a real variable is a rule which assigns a unique complex number to each real number x in the domain of f; thus for each x, $f(x) = u(x) + iv(x)$, where $u(x)$ and $v(x)$ are real numbers. A complex-valued function therefore consists of a pair of real functions, u and v, which are called respectively the *real part* of f and the *imaginary part* of f. Limits are defined for complex functions in the same way as for real functions, only replacing the notion of distance between real numbers by the distance between complex numbers. If $z = u + iv$ and $w = r + is$, with u, v, r, s real, then the *distance between z and w is*

$$|z - w| = \sqrt{(u - r)^2 + (v - s)^2}. \tag{1}$$

DEFINITION 1. If $f = u + iv$ is a complex function, then

$$\lim_{x \to x_0} f(x) = u_0 + iv_0$$

if and only if for each positive number ϵ, there is a positive number δ such that $|f(x) - (u_0 + iv_0)| = |u(x) + iv(x) - (u_0 + iv_0)| < \epsilon$ whenever $0 < |x - x_0| < \delta$.

From (1) it is clear that $|u(x) + iv(x) - (u_0 + iv_0)|$ is small if and only if both $|u(x) - u_0|$ and $|v(x) - v_0|$ are small. With this in mind, the following theorem follows readily from the definition above.

THEOREM 1. *If $f = u + iv$, then $\lim_{x \to x_0} f(x) = u_0 + iv_0$ if and only if $\lim_{x \to x_0} u(x) = u_0$ and $\lim_{x \to x_0} v(x) = v_0$.*

Proof. Problem 1.

The derivative of a complex-valued function of a real variable is defined in the same way as the derivative of a real function, and we use the same notation.

DEFINITION 2. If $f = u + iv$, then we define

$$Df(x) = f'(x) = \lim_{h \to 0} \frac{f(x + h) - f(x)}{h}.$$

83

The difference quotient for f can be written

$$\frac{f(x+h) - f(x)}{h} = \frac{u(x+h) + iv(x+h) - u(x) - iv(x)}{h}$$

$$= \frac{u(x+h) - u(x)}{h} + i\frac{v(x+h) - v(x)}{h}. \qquad (2)$$

Applying Theorem 1 to (2), we see that both expressions

$$\frac{u(x+h) - u(x)}{h} \qquad \text{and} \qquad \frac{v(x+h) - v(x)}{h}$$

must approach limits if f is differentiable. Therefore f is differentiable if and only if both u and v are, and we have the formula

$$Df(x) = f'(x) = u'(x) + iv'(x). \qquad (3)$$

The familiar theorems for derivatives of real functions go over unchanged to complex functions (Problem 2). The sum, product and quotient rules, and the chain rule, are exactly the same for real and complex functions. For example,

$$D(f(x)g(x)) = f(x)g'(x) + f'(x)g(x),$$

and

$$Df(g(x)) = f'(g(x))g'(x),$$

whether f and g are real or complex.

EXAMPLE 1.

(i) $D[x^2 + x + i(x^3 + \cos x)] = 2x + 1 + i(3x^2 - \sin x)$

(ii) $D[e^x (\sin x + i \cos x)] = e^x (\cos x - i \sin x) + e^x (\sin x + i \cos x)$
$$= e^x (\cos x + \sin x) + ie^x (\cos x - \sin x).$$

Here we used the product rule with $f(x) = e^x$, $g(x) = \sin x + i \cos x$.

(iii) $D(x + ix^2)^2 = 2(x + ix^2)(1 + i2x)$
$$= 2(x - 2x^3) + i6x^2.$$

Complex functions have been introduced so we will have at our disposal the complex exponential functions, e^{rx}, with r complex. These functions, defined next, are the basic solutions of the linear homogeneous equation.

DEFINITION 3. If a and b are real numbers, then

$$e^{(a+ib)x} = e^{ax} \cos bx + ie^{ax} \sin bx. \qquad (4)$$

The motivation for this definition is discussed in Problem 5. Let us now compute the derivative of $e^{(a+ib)x}$ in accordance with (3).

$$
\begin{aligned}
De^{(a+ib)x} &= D[e^{ax}\cos bx + ie^{ax}\sin bx] \\
&= ae^{ax}\cos bx - be^{ax}\sin bx + i(ae^{ax}\sin bx + be^{ax}\cos bx) \\
&= (a+ib)e^{ax}\cos bx + i(a+ib)e^{ax}\sin bx \\
&= (a+ib)[e^{ax}\cos bx + ie^{ax}\sin bx].
\end{aligned}
$$

Comparing this formula with (4) we see that

$$
De^{rx} = re^{rx}, \qquad \textit{whether r is real or complex.}^* \tag{5}
$$

Just as for real functions, we say a complex-valued function is a solution of a differential equation if the equation becomes an identity when the function and its derivatives are substituted. For example, if $y = e^{ix}$, then $y' = ie^{ix}$, $y'' = -e^{ix}$, and y is a solution of $y'' + y = 0$. We are of course primarily interested in real solutions, and the following theorem makes the connection between real and complex solutions of homogeneous linear equations.

THEOREM 2. *If $y = u + iv$, where u and v are real functions, then y is a solution of*

$$
y^{(n)} + p_{n-1}y^{(n-1)} + \cdots + p_1 y' + p_0 y = 0, \tag{6}
$$

where $p_0, \ldots, p_{n-1}$ are real functions or real constants if and only if u and v are solutions of (6).

Proof. The various derivatives of y are given by $y^{(k)} = u^{(k)} + iv^{(k)}$. Substituting y into the equation and separating the real and imaginary parts, we get the complex function

$$
\begin{aligned}
[u^{(n)} + p_{n-1}u^{(n-1)} &+ \cdots + p_1 u' + p_0 u] \\
&+ i[v^{(n)} + p_{n-1}v^{(n-1)} + \cdots + p_1 v' + p_0 v].
\end{aligned}
$$

This function is identically zero if and only if its real and imaginary parts are; i.e., if and only if both u and v are solutions of (6).

Returning to the example above, the fact that $e^{ix} = \cos x + i\sin x$ is a solution of $y'' + y = 0$ gives us the real solutions $\cos x$ and $\sin x$.

Now let us try to find solutions of the form $y = e^{rx}$ for any homogeneous linear equation with constant real coefficients.

$$
y^{(n)} + p_{n-1}y^{(n-1)} + \cdots p_1 y' + p_0 y = 0. \tag{7}
$$

---

* One way of *defining* the real exponential e^{ax} is to say that it is the solution of $y' = ay$ such that $y(0) = 1$. The fact that e^{rx}, with r complex, satisfies $y' = ry$, $y(0) = 1$ is one justification of the definition (4).

Substituting $y = e^{rx}$, r possibly complex, we get

$$e^{rx}(r^n + p_{n-1}r^{n-1} + \cdots + p_1 r + p_0) = 0, \qquad (8)$$

which is equivalent to

$$r^n + p_{n-1}r^{n-1} + \cdots + p_1 r + p_0 = 0. \qquad (9)$$

Thus e^{rx} is a solution of (7) if r is a real or complex root of (9). As in Section 3–4, (9) will be called the *auxiliary equation* for (7). If (9) has a complex root $a + ib$, then the conjugate complex number $a - ib$ will also be a root (Problem 6), since the coefficients p_i are real. For each pair of conjugate complex roots $a \pm ib$ of (9) we have the two complex solutions of (7), $e^{ax} \cos bx + ie^{ax} \sin bx$ and $e^{ax} \cos bx - ie^{ax} \sin bx$, and the two real solutions $e^{ax} \cos bx$ and $e^{ax} \sin bx$. For each real root r, we have the real solution e^{rx}.

Therefore, *if the auxiliary equation has no repeated roots, the required n real solutions are the exponential functions corresponding to the real roots, and the real and imaginary parts of the exponential functions corresponding to the complex roots.* The proof that the n functions found in this way are linearly independent, and the solution of (7) when the auxiliary equation has repeated roots are given in Section 4–3.

EXAMPLE 2. $y''' - 3y'' + 7y' - 5y = 2 - 5x - 8e^{-x}$.

For this nonhomogeneous equation, we must find one particular solution and the general solution of the reduced equation. The auxiliary equation is $r^3 - 3r^2 + 7r - 5 = 0$. By inspection, 1 is found to be a root of the auxiliary equation, and therefore $r - 1$ is a factor of the left side. Division by $r - 1$ gives the other factor $r^2 - 2r + 5$. The remaining two roots of the auxiliary equation are the roots of $r^2 - 2r + 5 = 0$, which are found by the quadratic formula to be $1 + 2i$ and $1 - 2i$. The complex function $e^{(1+2i)x}$ is a solution of the reduced equation, and $e^x \cos 2x$, $e^x \sin 2x$ are the corresponding real solutions. The general solution of the reduced equation is therefore

$$y = c_1 e^x + c_2 e^x \cos 2x + c_3 e^x \sin 2x.$$

To find a particular solution of the given equation, we try a function of the form $A + Bx$ to fit the term $2 - 5x$, and a function of the form Ce^{-x} to fit the term $-8e^{-x}$. Substitution of $y = A + Bx + Ce^{-x}$ gives $(7B - 5A) - 5Bx - 16Ce^{-x}$. Therefore $C = \frac{1}{2}$, $B = 1$, and $A = 1$. The general solution is

$$y = 1 + x + \tfrac{1}{2}e^{-x} + c_1 e^x + c_2 e^x \cos 2x + c_3 e^x \sin 2x.$$

PROBLEMS

1. Make a formal proof of Theorem 1 by filling in the details of this outline:
First, suppose $\lim_{x \to x_0} u(x) = u_0$ and $\lim_{x \to x_0} v(x) = v_0$. Let $\epsilon > 0$. Choose
positive numbers δ_1, δ_2 such that $|u(x) - u_0| < \epsilon/\sqrt{2}$ if $0 < |x - x_0| < \delta_1$
and $|v(x) - v_0| < \epsilon/\sqrt{2}$ if $0 < |x - x_0| < \delta_2$. Show that $|f(x) - (u_0 + iv_0| < \epsilon$
if $0 < |x - x_0| < \min(\delta_1, \delta_2)$. Second, suppose $\lim_{x \to x_0} f(x) = u_0 + iv_0$. Let
$\epsilon > 0$. Pick $\delta > 0$ such that $|f(x) - (u_0 + iv_0)| < \epsilon$ if $0 < |x - x_0| < \delta$.
Show that $|u(x) - u_0| < \epsilon$ and $|v(x) - v_0| < \epsilon$ if $0 < |x - x_0| < \delta$.

2. Let f and g be differentiable complex functions with $f = u + iv$ and
$g = r + is$. Use the differentiation formulas for real functions and (3) to verify
the following:

   (a) $D(f + g) = Df + Dg$ (b) $D(cf) = cDf$, c a complex constant.
   (c) $D(fg) = fDg + gDf$ (d) $D(f/g) = (gDf - fDg)/g^2$

Note. Since the real functions are a subclass of the complex functions (they are
the complex functions $u + iv$ with $v \equiv 0$), these formulas hold if one function is
real and the other is complex.

3. (a) Differentiate $e^x \sin x + ie^x \cos x$ and check the result which was obtained in Example 1(ii) with the product rule.

   (b) Expand $(x + ix^2)^2$ and then differentiate to check the chain rule
computation of Example 1(iii).

4. Compute the following derivatives.

   (a) $D(x \sin x + i(x^2 + 1))$ (b) $D(x^2 + 2ix + 1 - ix^2)$
   (c) $D\sqrt{-1 - x^2}$ (d) $D[(x + ix^2)(\sin x + i \cos x)]$

5. Show that the substitution of ib for x in the series for e^x [(6), Section 2–6]
gives $e^{ib} = \cos b + i \sin b$. The property $e^p e^q = e^{p+q}$ is an algebraic property
of the series for e^p, e^q, and e^{p+q} and doesn't depend on whether p and q are real
or complex. This shows that $e^{a+ib} = e^a e^{ib} = e^a(\cos b + i \sin b)$, and $e^{(a+ib)x} = e^{ax+ibx} = e^{ax}(\cos bx + i \sin bx)$.

6. If $z = a + ib$, with a and b real, then the conjugate of z is the number
$\bar{z} = a - ib$. Let $z = a + ib$, and $w = c + id$ be any complex numbers.

   (a) Verify that $\overline{z + w} = \bar{z} + \bar{w}$.   (b) Verify that $\overline{(zw)} = \bar{z} \bar{w}$.

   (c) Let $P(z) = p_0 + p_1 z + \cdots + p_n z^n$, with $p_0, \ldots, p_n$ real. Show that
$P(\bar{z}) = \overline{P(z)}$.

   (d) Show that if z_0 is a root of the polynomial equation $P(z) = 0$, P as in (c),
then $\bar{z}_0$ is also a root.

7. Solve the following equations.

   (a) $y^{(iv)} - y = 4 + e^{2x}$ (b) $y''' + 3y'' + 4y' + 2y = 2x$
   (c) $y''' + 4y'' + 5y' = 5$
   (d) $y''' - 6y'' + 13y' - 10y = 10x^2 - 6x - 4$

8. (a) Verify that $2 + 2i$ is a root of the equation

$$r^4 - 5r^3 + 40r - 96 = 0.$$

   (b) Solve the equation $y^{(iv)} - 5y''' + 40y' - 96y = 0.$

ANSWERS

4. (a) $x \cos x + \sin x + i(2x)$

    (b) $2x + i(2 - 2x)$

    (c) $ix(1 + x^2)^{-1/2}$

    (d) $-x \cos x + (x^2 + 1) \sin x + i[(x^2 + 1) \cos x + x \sin x]$

7. (a) $y = -4 + \frac{1}{15}e^{2x} + c_1 e^x + c_2 e^{-x} + c_3 \cos x + c_4 \sin x$

    (b) $y = x - 2 + c_1 e^{-x} + c_2 e^{-x} \cos x + c_3 e^{-x} \sin x$

    (c) $y = x + c_1 + c_2 e^{-2x} \cos x + c_3 e^{-2x} \sin x$

    (d) $y = -(x + 1)^2 + e^{2x}(c_1 + c_2 \cos x + c_3 \sin x)$

8. $y = c_1 e^{-3x} + c_2 e^{4x} + c_3 e^{2x} \cos 2x + c_4 e^{2x} \sin 2x$

4–2 Linear differential operators. An *operator* is a rule which associates a unique function with each function in some set. An operator is therefore itself a function, which is defined on some set of functions rather than numbers, and whose values are functions. An operator A, like any other function, can be defined by prescribing the value Ay of the operator at an arbitrary function y. Thus the formula $Ay = y' + 3y$ defines the operator A just as the formula $z(x) = x^2 + 2$ defines the function z. For the A and z just defined, for example, $(Az)(x) = 2x + 3(x^2 + 2)$.

The operators of interest to us here, generalizations of the familiar derivative operators D, D^2, etc., are those of the form

$$Ly = p_n y^{(n)} + p_{n-1} y^{(n-1)} + \cdots + p_1 y' + p_0 y \tag{1}$$

or, equivalently,

$$Ly = p_n D^n y + p_{n-1} D^{n-1} y + \cdots + p_1 Dy + p_0 y, \tag{1}$$

where $p_0, p_1, \ldots, p_n$ are constants. The operator L defined by (1) will also be denoted by $L(D)$ or by a formal polynomial in D,

$$L = L(D) = p_n D^n + p_{n-1} D^{n-1} + \cdots + p_1 D + p_0. \tag{2}$$

An operator of the form (2), as defined by (1), is called a *linear differential operator*. The general linear differential equation, with constant coefficients, can be written in operator notation as follows:

$$Ly = L(D)y = q \qquad (p_n = 1), \tag{3}$$

or

$$(D^n + p_{n-1} D^{n-1} + \cdots + p_1 D + p_0)y = q. \tag{4}$$

The linearity properties of the operators (1) are expressed in the following theorem.

THEOREM 1. *If L is a linear differential operator, then $L(y_1 + y_2) = Ly_1 + Ly_2$, and $L(cy) = cLy$ for all functions y, y_1, y_2 and all constants c.*

Proof. The first property has already been proved, with different notation, in Section 3–3 (the superposition principle). To show that $L(cy) = cLy$, note that $p_k D^k(cy) = cp_k D^k y$, so c can be factored from each term in $L(cy)$ to give cLy.

The advantage of the expression (2) lies in the fact that the ordinary multiplication and addition of polynomials corresponds to the natural operator multiplication and addition defined below. This allows us to factor the polynomial (2) and so replace a homogeneous linear equation with equations of smaller order. Although we are interested in equations (4) with real coefficients, we may need to allow complex coefficients to factor the operator (2). Since the rules for differentiation are the same for real or complex functions and constants, it causes no difficulty to allow the p_i in (2) to be complex and to admit complex-valued functions in the definition (1).

For any operators A and B, we define new operators $A + B$ and AB by

$$(A + B)y = Ay + By, \tag{5}$$

$$(AB)y = A(By). \tag{6}$$

For example, if $Ay = y' + 3y$ and $By = xy$, then $(A + B)y = y' + 3y + xy$, and $(AB)y = A(xy) = (xy)' + 3(xy) = xy' + y + 3xy$. Note that $AB \neq BA$ in general; here, for instance, $(BA)y = B(y' + 3y) = xy' + 3xy \neq (AB)y$.

THEOREM 2. *If L_1 and L_2 are linear differential operators, then so is $L_1 + L_2$ [defined in (5)] and the polynomial form (2) of $L_1 + L_2$ is the ordinary formal sum of the polynomials for L_1 and L_2.*

Proof. If

$$L_1 = p_n D^n + \cdots + p_1 D + p_0,$$
$$L_2 = q_m D^m + \cdots + q_1 D + q_0 \qquad \text{(say } m < n\text{)},$$

then

$$\begin{aligned}
(L_1 + L_2)y &= L_1 y + L_2 y \\
&= p_n D^n y + \cdots + p_1 D y + p_0 y \\
&\quad + q_m D^m y + \cdots + q_1 D y + q_0 y \\
&= p_n D^n y + \cdots + (p_m + q_m)D^m y + \cdots \\
&\qquad + (p_1 + q_1)D y + (p_0 + q_0)y.
\end{aligned}$$

This shows that $L_1 + L_2$ has the form (1) and is therefore a linear differential operator. Moreover, the polynomial form (2) of $L_1 + L_2$ is clearly the sum of the polynomials for L_1 and L_2.

Remark. The right side of (2) was introduced as a formal expression, with the plus signs serving only to make spaces between the terms. By virtue of Theorem 2 we can now regard the right side of (2) as a genuine sum, as defined in (5), of the operators p_0, $p_1 D$, etc., and this is consistent with the definition (1).

Recall that the coefficients in a linear operator are always assumed to be constants. Theorem 2 would still hold if the coefficients $p_0, p_1, \ldots, p_n$ were allowed to be functions, but Theorem 3 would definitely not be true for nonconstant coefficients.

THEOREM 3. *If L_1 and L_2 are linear differential operators, then so is $L_1 L_2$ and the polynomial form of $L_1 L_2$ is the ordinary product of the polynomials for L_1 and L_2.*

Proof. If L_1 and L_2 are as in Theorem 2, then

$$(L_1 L_2)y = L_1(L_2 y)$$

$$= L_1\left(\sum_{j=0}^{m} q_j D^j y\right)$$

$$= \sum_{i=0}^{n} p_i D^i \left(\sum_{j=0}^{m} q_j D^j y\right)$$

$$= \sum_{i=0}^{n} p_i \sum_{j=0}^{m} q_j D^{i+j} y$$

$$= \sum_{i=0}^{n} \sum_{j=0}^{m} p_i q_j D^{i+j} y.$$

The right side above is of the form (1), showing that $L_1 L_2$ is a linear differential operator. Also, the polynomial form of $L_1 L_2$ is clearly the formal product

$$\left(\sum_{i=0}^{n} p_i D^i\right)\left(\sum_{j=0}^{m} q_j D^j\right) = \sum_{i=0}^{n} \sum_{j=0}^{m} p_i q_j D^{i+j}$$

of the polynomials for L_1 and L_2.

COROLLARY. *If L_1 and L_2 are linear differential operators, then*

$$L_1 L_2 = L_2 L_1.$$

Proof. Problem 3(c).

The utility of the results above is illustrated in the following example.

EXAMPLE 1. Consider the homogeneous linear equation

$$y''' - y'' + y' - y = 0. \tag{7}$$

This can be written in operator notation as

$$(D^3 - D^2 + D - 1)y = 0. \tag{8}$$

The polynomial $D^3 - D^2 + D - 1$ can be factored as

$$D^3 - D^2 + D - 1 = (D^2 + 1)(D - 1) = (D + i)(D - i)(D - 1).$$

Because of Theorem 3, we can write (8) in any of the equivalent forms:

$$\begin{aligned} (D^2 + 1)[(D - 1)y] &= 0, \\ (D - 1)(D + i)[(D - i)y] &= 0, \\ (D - 1)(D - i)[(D + i)y] &= 0. \end{aligned} \tag{9}$$

For any linear operator L, it is clear from (1) that $L0 = 0$, where "0" denotes the zero function. It follows that y is a solution of (7), (8), or (9) if y is a solution of any of the equations

$$\begin{aligned} (D - 1)y &= 0, \\ (D - i)y &= 0, \\ (D + i)y &= 0. \end{aligned} \tag{10}$$

Thus the third order homogeneous equation (7) is reduced to the three first order equations (10), whose solutions are e^x, $e^{ix} = \cos x + i \sin x$, and $e^{-ix} = \cos x - i \sin x$. From Theorem 2, Section 4–1, we know that the complex functions are solutions of (7) only if their real and imaginary parts are. So the required three real solutions of (7) are e^x, $\sin x$, and $\cos x$.

EXAMPLE 2. $(D^3 - 2D^2 + 2D)y = 0$.

This equation can be written in either of the forms

$$D[(D^2 - 2D + 2)y] = 0 \quad \text{or} \quad (D^2 - 2D + 2)[Dy] = 0.$$

The solutions will be the solutions of $Dy = 0$, plus the solutions of $(D^2 - 2D + 2)y = 0$. Instead of factoring the quadratic operator into linear factors, $D^2 - 2D + 2 = [D - (1 + i)][D - (1 - i)]$, as in Example 1, we can use the method of Section 4–1. The roots of the auxiliary equation are $1 \pm i$, giving the real solutions $e^x \cos x$ and $e^x \sin x$.

There is no real difference in the calculations involved in the methods of Section 4–1 and the operator methods of this section as applied to homogeneous linear equations with constant coefficients. Factoring the polynomial operator certainly amounts to the same thing as solving the auxiliary equation. The operator methods, however, are the more flexible and powerful. We will treat the case of repeated roots of the auxiliary equation in the next section with operator methods. The next example illustrates the use of operators to solve a nonhomogeneous equation.

EXAMPLE 3. $y'' - y' - 2y = xe^x$.

Let us write the equation in the form

$$(D - 2)[(D + 1)y] = xe^x. \tag{11}$$

Clearly y will be a solution of (11) if and only if y is a solution of

$$(D + 1)y = u, \tag{12}$$

where u is a solution of

$$(D - 2)u = xe^x. \tag{13}$$

That is, we can replace the second order linear equations (11) by the pair of simultaneous linear first order equations (12) and (13). Multiplying (13) by the integrating factor e^{-2x}, we get

$$e^{-2x}u' - 2ue^{-2x} = xe^{-x},$$

and hence

$$ue^{-2x} = \int xe^{-x}\, dx + c_1.$$

Integration by parts gives

$$\int xe^{-x}\, dx = -xe^{-x} + \int e^{-x}\, dx$$
$$= -e^{-x}(x + 1),$$

and therefore the solutions u of (13) are

$$u = -e^x(x + 1) + c_1 e^{2x}. \tag{14}$$

From (12) and (14), we see that y is a solution of (11) if and only if y is a solution of

$$(D + 1)y = -e^x(x + 1) + c_1 e^{2x}.$$

This linear first order equation has the integrating factor e^x, and we get

$$e^x y' + e^x y = -e^{2x}(x + 1) + c_1 e^{3x}.$$

The solutions are given by

$$e^x y = -\int e^{2x}(x + 1)\, dx + c_1 e^{3x} + c_2,$$

where $c_1/3$ has been replaced by c_1. Integration by parts gives

$$e^x y = -\tfrac{1}{2}e^{2x}(x + 1) + \int \tfrac{1}{2}e^{2x}\, dx + c_1 e^{3x} + c_2$$
$$= -\tfrac{1}{2}e^{2x}(x + 1) + \tfrac{1}{4}e^{2x} + c_1 e^{3x} + c_2.$$

The general solution of (11) is, accordingly,

$$y = -\tfrac{1}{4}e^x(2x + 1) + c_1 e^{2x} + c_2 e^{-x}.$$

Note that the solutions $c_1 e^{2x} + c_2 e^{-x}$ of the reduced equation appear automatically, as they must since the procedure finds *all* solutions of (11).

EXAMPLE 4. $(D - r)^2 y = 0$ (r is real.)

In Section 3–4 we introduced the solution of this equation and verified that it worked. Let us now show how the method of Example 3 can be used to derive the solution. The equation is equivalent to the system

$$(D - r)u = 0, \qquad (D - r)y = u.$$

The solutions of the first equation are

$$u = c_1 e^{rx},$$

and hence the second equation becomes

$$(D - r)y = c_1 e^{rx}.$$

Multiplying by the integrating factor e^{-rx}, we get

$$e^{-rx}y' - re^{-rx}y = c_1,$$
$$ye^{-rx} = c_1 x + c_2,$$
$$y = e^{rx}(c_1 x + c_2).$$

PROBLEMS

1. Let A and B be the operators defined by $Ay = y^2 + xy$, $By = 2y'$. Compute the following:

(a) Ax^2 (b) Bx^2 (c) $(A + B)x^2$
(d) $A(Bx^2)$ (e) $B(Ax^2)$

2. Verify the superposition principle, $L(y_1 + y_2) = Ly_1 + Ly_2$, for linear differential operators L, using the polynomial notation for L (cf. Theorem 3, Section 3-3).

3. (a) Prove that operator addition (5) is commutative and associative: $A + B = B + A$, and $(A + B) + C = A + (B + C)$ for all operators A, B, C.
   (b) Prove that operator multiplication is associative: $(AB)C = A(BC)$ for all A, B, C.
   (c) Prove the corollary to Theorem 3.

4. Solve the following equations by the method of Example 1.
   (a) $(D^3 - 6D^2 + 5D)y = 0$ (b) $y''' + 2y'' - y' - 2y = 0$
   (c) $(D^2 - 1)(D^2 + 1)y = 0$ (d) $(D^2 - 2D + 2)(D - 1)y = 0$

5. Show that if y is a solution of $(D^2 + 1)y = e^{2x}$, then y is a solution of $(D - 2)(D^2 + 1)y = 0$. Solve both equations.

6. Suppose that L is a linear differential operator and $L(\cos x) \neq 0$, $L(\sin x) \neq 0$. Prove that the equation $Ly = \cos x$ has a solution of the form $A \cos x + B \sin x$.

7. Write an equivalent system of first order equations, as in Example 3, and solve $(D - 1)^2 y = e^x$.

8. Write as a system and solve $(D + 1)(D - 1)y = x$.

9. Solve $(D - a)^3 y = 0$ (a real) by changing to a system of three first order equations.

10. Recall from Section 3-4 that $(D - a)^2 x e^{ax} = 0$ (a real).
   (a) Show that if y is a solution of (i) $(D - 1)(D - 2)y = 2xe^x$, then y is a solution of (ii) $(D - 2)(D - 1)^3 y = 0$.
   (b) Find all solutions of (ii) (see Problem 9).
   (c) Find all solutions of the reduced form of (i).
   (d) Find a particular solution of (i) by the method of undetermined coefficients. Note that the form of your trial solution is determined by your answers to (a), (b), and (c).

ANSWERS

1. (a) $x^4 + x^3$ (b) $4x$ (c) $x^4 + x^3 + 4x$
   (d) $20x^2$ (e) $8x^3 + 6x^2$

4. (a) $y = c_1 + c_2 e^x + c_3 e^{5x}$
   (b) $y = c_1 e^x + c_2 e^{-x} + c_3 e^{-2x}$
   (c) $y = c_1 e^x + c_2 e^{-x} + c_3 \cos x + c_4 \sin x$
   (d) $y = c_1 e^x + c_2 e^x \cos x + c_3 e^x \sin x$

5. $y = c_1 e^{2x} + c_2 \cos x + c_3 \sin x$
   $y = \frac{1}{5}e^{2x} + c_1 \cos x + c_2 \sin x$

7. $y = \frac{1}{2}x^2 e^x + e^x(c_1 x + c_2)$

8. $y = -x + c_1 e^{-x} + c_2 e^x$

9. $y = e^{ax}(c_1 x^2 + c_2 x + c_3)$

10. (a) $y = c_1 e^{2x} + e^x(c_2 x^2 + c_3 x + c_4)$
    (c) $y = c_1 e^{2x} + c_2 e^x$
    (d) $y = (-x^2 - 2x)e^x$

4–3 Homogeneous equations with constant coefficients. With the methods now at our disposal we can finish the discussion of the linear homogeneous equation with constant real coefficients

$$(D^n + p_{n-1}D^{n-1} + \cdots + p_1 D + p_0)y = 0. \tag{1}$$

First let us recall from the preceding sections some general facts about (1).

We know (Theorem 7, Section 3–3) that (1) has n linearly independent solutions, and if $y_1, \ldots, y_n$ are any n linearly independent solutions, then the family

$$c_1 y_1 + \cdots + c_n y_n \tag{2}$$

is the set of all solutions of (1).

If $f = u + iv$ is any complex valued solution of (1), then (Theorem 2, Section 4–1) u and v are solutions of (1).

The operator $L = D^n + \cdots + p_1 D + p_0$ in (1) can be factored into linear factors

$$L = (D - r_1)^{k_1}(D - r_2)^{k_2} \cdots (D - r_s)^{k_s}, \tag{3}$$

where $r_1, r_2, \ldots, r_s$ are distinct numbers. Some of the numbers r_j may be complex, and if so, the complex r_j occur in conjugate pairs (Problem 6, Section 4–1). All solutions of the equations

$$(D - r_j)^{k_j}y = 0 \tag{4}$$

are (possibly complex) solutions of (1).

The attack on (3) will proceed as follows. If r_j is real, we will find k_j real solutions of (4) which are necessarily also solutions of (1). If r_j is complex, then we find $2k_j$ real solutions of

$$(D - r_j)^{k_j}(D - \bar{r}_j)^{k_j}y = 0 \tag{5}$$

which are also solutions of (1). In this way we accumulate n real solutions of (1). These solutions are linearly independent for any numbers $r_1, \ldots, r_s$, and their linear combinations therefore constitute the general solution of (1).

THEOREM 1. *If m is a positive integer and r is a real or complex number, then*

$$(D - r)^{m+1}x^m e^{rx} = 0. \tag{6}$$

Proof. Calculating according to (5), Section 4–1, and the product rule, we see that

$$\begin{aligned}
(D - r)x^m e^{rx} &= D(x^m e^{rx}) - rx^m e^{rx} \\
&= mx^{m-1}e^{rx} + rx^m e^{rx} - rx^m e^{rx} \\
&= mx^{m-1}e^{rx}.
\end{aligned} \tag{7}$$

Using (7) again with m replaced by $m - 1$ gives

$$\begin{aligned}
(D - r)^2 x^m e^{rx} &= (D - r)mx^{m-1}e^{rx} \\
&= m(D - r)x^{m-1}e^{rx} \\
&= m(m - 1)x^{m-2}e^{rx}.
\end{aligned} \tag{8}$$

It is clear that we can repeat this process as long as the exponent of x is positive, and after m steps, we get

$$(D - r)^m x^m e^{rx} = m(m - 1) \cdots 2 \cdot 1 e^{rx}. \tag{9}$$

Hence,

$$(D - r)^{m+1} x^m e^{rx} = m!(D - r)e^{rx} = 0. \tag{10}$$

COROLLARY 1. *If p is any of the numbers $0, 1, 2, \ldots, k - 1$, then* $(D - r)^k x^p e^{rx} = 0.$

Proof. Problem 1.

COROLLARY 2. *If P is any polynomial of degree $k - 1$ or less, then* $(D - r)^k P(x)e^{rx} = 0.$

Proof. Problem 1.

From Corollary 1, we see that if r is real, then

$$e^{rx}, xe^{rx}, \ldots, x^{k-1}e^{rx} \tag{11}$$

is a set of k real solutions of

$$(D - r)^k y = 0. \tag{12}$$

If $(D - r)^k$ is a factor of the operator (3), then the solutions (11) of (12) are also solutions of (1). Thus we get k solutions of (1) corresponding to each linear factor of multiplicity k in the operator.

Now suppose r is complex, $r = a + ib$, and $(D - r)^k$ is a factor of (3). Then $(D - \bar{r})^k$ is also a factor ($\bar{r} = a - ib$). The equation

$$(D - r)^k(D - \bar{r})^k y = 0 \tag{13}$$

is a linear equation of order $2k$ with *real* coefficients (Problem 3). From Corollary 1, this equation has the complex solutions

$$e^{rx}, xe^{rx}, \ldots, x^{k-1}e^{rx} \tag{14}$$

and also the solutions

$$e^{\bar{r}x}, xe^{\bar{r}x}, \ldots, x^{k-1}e^{\bar{r}x}. \tag{15}$$

The real and imaginary parts of the functions (14) are the same, apart from sign, as the real and imaginary parts of the functions (15); they are the functions

$$e^{ax} \cos bx, \ xe^{ax} \cos bx, \ldots, \ x^{k-1}e^{ax} \cos bx,$$
$$e^{ax} \sin bx, \ xe^{ax} \sin bx, \ldots, \ x^{k-1}e^{ax} \sin bx. \tag{16}$$

These solutions of (13) are of course also solutions of (1), and we have found $2k$ real solutions of (1) corresponding to each pair of linear factors $(D - r)^k(D - \bar{r})^k$ in the operator (3). Altogether we have found n real solutions of the nth order equation (1).

THEOREM 2. *Let* $Ly = 0$ *be an nth order linear equation with constant real coefficients. Let*

$$L = (D - r_1)^{k_1}(D - r_2)^{k_2} \ldots (D - r_s)^{k_s},$$

where $r_1, \ldots, r_s$ *are distinct numbers.*
If r_j *is real, then* $Ly = 0$ *has the solutions*

$$e^{r_j x}, \ xe^{r_j x}, \ldots, \ x^{k_j - 1}e^{r_j x}.$$

If $r_j = a_j + ib_j$, *then* $Ly = 0$ *has the solutions*

$$e^{a_j x} \cos b_j x, \ xe^{a_j x} \cos b_j x, \ldots, \ x^{k_j - 1}e^{a_j x} \cos b_j x,$$
$$e^{a_j x} \sin b_j x, \ xe^{a_j x} \sin b_j x, \ldots, \ x^{k_j - 1}e^{a_j x} \sin b_j x.$$

The n solutions of $Ly = 0$ *given above are linearly independent for any operator L, and the general solution of* $Ly = 0$ *is the set of all linear combinations of these n solutions.*

Proof. The only statement in the theorem which has not yet been verified is the linear independence of the given solutions for any equation. The proof of linear independence in general involves such complicated notation that it becomes uninstructive, and we will prove the statement only in the case where the operator can be factored into *real* linear factors,

$$L = (D - a_1)^{k_1}(D - a_2)^{k_2} \ldots (D - a_s)^{k_s}$$

with $a_1, \ldots, a_s$ distinct real numbers. The linear combinations of the solutions found for $Ly = 0$ can be written

$$P_1(x)e^{a_1 x} + P_2(x)e^{a_2 x} + \ldots + P_s(x)e^{a_s x}, \tag{17}$$

where $P_1, \ldots, P_s$ are arbitrary polynomials with respective degrees

$k_1 - 1, \ldots, k_s - 1$. Let a_1 be the largest of the distinct numbers a_1, $\ldots, a_s$, so that $a_j - a_1 < 0$ for $j = 2, \ldots, s$, and $e^{(a_j - a_1)x}$ is a decreasing function. Suppose that (17) is identically zero for some polynomials $P_1, \ldots, P_s$. We must show that all the coefficients in each polynomial are zero, which is the same as showing that each polynomial is the zero function. If (17) is identically zero, then multiplying by $e^{-a_1 x}$ and transposing the first term, we get

$$-P_1(x) \equiv P_2(x)e^{(a_2 - a_1)x} + \ldots + P_s(x)e^{(a_s - a_1)x}. \tag{18}$$

For any exponential e^{qx}, $q < 0$, and any polynomial $R(x)$, we have (Problem 4)

$$\lim_{x \to \infty} R(x)e^{qx} = 0.$$

This implies, in view of (18), that $\lim_{x \to \infty} P_1(x) = 0$. However, the only polynomial which tends to zero as $x \to \infty$ is the zero polynomial (Problem 5), so $P_1(x) \equiv 0$. Having shown that $P_1(x) \equiv 0$, the assumption that (17) is identically zero becomes

$$P_2(x)e^{a_2 x} + \cdots + P_s(x)e^{a_s x} \equiv 0.$$

The same argument as above can now be used to show that $P_2(x) \equiv 0$. Continuing in this way, we see that (17) is identically zero only if all the coefficients are zero, and the solutions found are linearly independent.

EXAMPLE 1.

(A) $(D - 1)^2(D^2 - 2D + 2)y$
$$= (D - 1)^2[D - (1 + i)][D - (1 - i)]y = 0,$$
$$y = c_1 e^x + c_2 x e^x + c_3 e^x \cos x + c_4 e^x \sin x.$$

(B) $(D + 2)(D^2 - 4D + 13)^2 y$
$$= (D + 2)[D - (2 + 3i)]^2[D - (2 - 3i)]^2 y = 0,$$
$$y = c_1 e^{-2x} + c_2 e^{2x} \cos 3x + c_3 x e^{2x} \cos 3x$$
$$+ c_4 e^{2x} \sin 3x + c_5 x e^{2x} \sin 3x$$
$$= c_1 e^{-2x} + e^{2x}[(c_2 + c_3 x) \cos 3x + (c_4 + c_5 x) \sin 3x].$$

(C) $(D^2 + 4)^2 D^2(D - 1)y = (D + 2i)^2(D - 2i)^2 D^2(D - 1)y = 0,$
$$y = (c_1 + c_2 x) \cos 2x + (c_3 + c_4 x) \sin 2x$$
$$+ c_5 + c_6 x + c_7 e^x.$$

EXAMPLE 2. Let us show that the solutions e^x, xe^x, $e^x \cos x$, $e^x \sin x$ of Example 1(A) are linearly independent. We could use the Wronskian test (corollary to Theorem 5, Section 3–3), but the computations involved are uninviting. Instead, we will argue directly from the definition. Suppose that for some numbers c_1, c_2, c_3, c_4, we have

$$c_1 e^x + c_2 x e^x + c_3 e^x \cos x + c_4 e^x \sin x \equiv 0.$$

Then

$$c_1 + c_2 x + c_3 \cos x + c_4 \sin x \equiv 0.$$

For $x = 0, 2\pi, 4\pi$, etc., this becomes

$$c_1 + c_2 x + c_3 \cos x + c_4 \sin x = (c_1 + c_3) + c_2 x = 0.$$

The linear function $(c_1 + c_3) + c_2 x$ is zero for more than one x only if it is identically zero, so $c_1 + c_3 = 0$, and $c_2 = 0$. For $x = \pi/2, \pi/2 + 2\pi$, etc.,

$$c_1 + 0x + c_3 \cos x + c_4 \sin x = c_1 + c_4 = 0.$$

From $c_1 + c_3 = 0$, $c_1 + c_4 = 0$, we get $c_3 = c_4$, and the original assumption becomes

$$c_1 + c_3 (\cos x + \sin x) \equiv 0.$$

This implies $c_1 = c_3 = 0$. Therefore, $c_1 = c_2 = c_3 = c_4 = 0$, and the solutions are linearly independent.

PROBLEMS

1. Prove the corollaries of Theorem 1.

2. Are the functions e^x, $(x - 1)e^x$, $(x^2 - x)e^x$ solutions of $(D - 1)^3 y = 0$? Is this set of functions linearly independent?

3. Show that $(D - r)^k (D - \bar{r})^k y = [(D - r)(D - \bar{r})]^k y = 0$ is a linear equation with real coefficients.

4. Let $q < 0$. Use l'Hospital's rule to show that $\lim_{x \to \infty} x e^{qx} = 0$. [*Hint:* $x e^{qx} = x/e^{-qx}$]. Prove by induction that $\lim_{x \to \infty} x^n e^{qx} = 0$ for any positive integer n. Show that $\lim_{x \to \infty} R(x) e^{qx} = 0$ for any polynomial R.

5. Show that if R is any nonconstant polynomial, then $\lim_{x \to \infty} |R(x)| = \infty$. [*Hint:* If $x > 0$, then

$$|p_0 + p_1 x + \cdots + p_n x^n| = x^n \left| \frac{p_0}{x^n} + \frac{p_1}{x^{n-1}} + \cdots + \frac{p_{n-1}}{x} + p_n \right|.$$

Show that if x is sufficiently large and $p_n \neq 0$, then the second factor on the right is greater than $|p_n|/2$, and $|R(x)| > \frac{1}{2}|p_n|x^n$.]

6. Solve the following equations.

(a) $(D - 1)^2(D^2 - 2D + 2)y = 0$
(b) $(D^2 + 1)(D^2 + 4D + 5)^2y = 0$
(c) $(D^2 - 1)(D^2 + 5D + 4)^2y = 0$
(d) $(D^2 - 4)(D + 2)(D^2 + 4)y = 0$

(e) $(D^6 - 1)y = 0$ (f) $(D^3 + D^2 + \frac{1}{2}D)y = 0$
(g) $(D^4 - D^2)y = 0$ (h) $(D^4 + 2D^3 + 2D^2 + 2D + 1)y = 0$

7. Show that the solutions are linearly independent.

(a) solutions of Problem 6(a)   (b) solutions of Problem 6(b)

8. Suppose y_0 is a solution of $Ly = 0$, where L is a linear differential operator (with constant coefficients). Prove that the following functions are necessarily also solutions of $Ly = 0$ or give an example (of L and y_0) to show they are not.

(a) $y_0'' - 2$ (b) $3y_0' + 2y_0$ (c) xy_0

Answers

6. (a) $y = (c_1 + c_2x)e^x + c_3e^x \cos x + c_4e^x \sin x$
   (b) $y = c_1 \cos x + c_2 \sin x + e^{-2x}[(c_3 + c_4x) \cos x + (c_5 + c_6x) \sin x]$
   (c) $y = c_1e^x + (c_2 + c_3x + c_4x^2)e^{-x} + (c_5 + c_6x)e^{-4x}$
   (d) $y = c_1e^{2x} + (c_2 + c_3x)e^{-2x} + c_4 \cos 2x + c_5 \sin 2x$
   (e) $y = c_1e^x + c_2e^{-x} + e^{(1/2)x}\left(c_3 \cos \frac{\sqrt{3}}{2} x + c_4 \sin \frac{\sqrt{3}}{2} x\right)$
   $$+ e^{-(1/2)x}\left(c_5 \cos \frac{\sqrt{3}}{2} x + c_5 \sin \frac{\sqrt{3}}{2} x\right)$$
   (f) $y = c_1 + e^{-(1/2)x}\left(c_2 \cos \frac{x}{2} + c_3 \sin \frac{x}{2}\right)$
   (g) $y = c_1 + c_2x + c_3e^x + c_4e^{-x}$
   (h) $y = (c_1 + c_2x)e^{-x} + c_3 \cos x + c_4 \sin x$

8. (b) $L(3D + 2)y_0 = (3D + 2)Ly_0 = (3D + 2)0$
   (a) and (c) are not necessarily also solutions.

4-4 Method of undetermined coefficients. In the preceding section we saw that the solutions of

$$(D - a)^ky = 0 \qquad (a \text{ real}) \tag{1}$$

and

$$(D - r)^k(D - \bar{r})^ky = 0 \qquad (r = a + ib) \tag{2}$$

are the functions

$$P(x)e^{ax} \cos bx + Q(x)e^{ax} \sin bx, \tag{3}$$

where P and Q are any polynomials of degree $k - 1$ and $b = 0$ in case (1). Since any linear operator can be factored into operators such as those in (1) or (2), it follows that all solutions of a linear equation $Ly = 0$ with constant coefficients are sums of functions of the form (3). Now look at these facts in this way: For any function q which is of the form (3), there is a linear operator L such that $Lq = 0$. This observation is the central idea in the following theorem.

THEOREM 1. *If L is a linear differential operator and P, Q are any given polynomials, then the equation*

$$Ly = P(x)e^{ax} \cos bx + Q(x)e^{ax} \sin bx \qquad (4)$$

has a particular solution y_0 which can be written

$$y_0 = P^*(x)e^{ax} \cos bx + Q^*(x)e^{ax} \sin bx \qquad (5)$$

for some polynomials P^* and Q^*.

Proof. We give the proof for the case $b \neq 0$, and ask the student to supply the proof for the case $b = 0$ (Problem 10).

Let $k - 1$ be the larger of the degrees of P and Q. Then by Corollary 2 of Theorem 1, Section 4–3,

$$[D - (a + ib)]^k[D - (a - ib)]^k(P(x)e^{ax} \cos bx + Q(x)e^{ax} \sin bx) = 0.$$

It follows that every solution of (4) is a solution of

$$[D - (a + ib)]^k[D - (a - ib)]^k Ly = 0. \qquad (6)$$

We know what all the solutions of (6) are, so we know what form a solution of (4) must take. Since we are looking for a particular solution of (4), we can disregard solutions of (6) which are solutions of the reduced form of (4) (i.e., of $Ly = 0$). The solutions of (6) which are not solutions of $Ly = 0$ are among the solutions corresponding to the factors

$$[D - (a + ib)]^{k+j}[D - (a - ib)]^{k+j}, \qquad (7)$$

where we assume that $[D - (a + ib)]$ occurs in L with multiplicity j (j possibly zero). The solutions of (6) corresponding to the factors (7) are the functions

$$P^*(x)e^{ax} \cos bx + Q^*(x)e^{ax} \sin bx, \qquad (8)$$

where P^* and Q^* are arbitrary polynomials of degree $k + j - 1$. Since all solutions of (4) which are not solutions of $Ly = 0$ appear among the functions (8), there will be some specific polynomials P^* and Q^* for which (8) is a solution of (4). If $[D - (a + ib)]$ does occur as a factor of L (i.e., $j \neq 0$), then some of the terms in (8) will be solutions of $Ly = 0$.

We can ignore these and take P^* and Q^* of the form

$$P^*(x) = A_j x^j + A_{j+1} x^{j+1} + \cdots + A_{j+k-1} x^{j+k-1},$$
$$Q^*(x) = B_j x^j + B_{j+1} x^{j+1} + \cdots + B_{j+k-1} x^{j+k-1}.$$

EXAMPLE 1.

$$(D^3 - D^2 + D - 1)y = (D - 1)(D^2 + 1)y = xe^x. \qquad (9)$$

Since $(D - 1)^2 x e^x = 0$, all solutions of (9) are solutions of

$$(D - 1)^3 (D^2 + 1)y = 0.$$

Some particular solution of (9) therefore must be a solution of $(D - 1)^3 y = 0$. The solutions of this, omitting ce^x which is a solution of the reduced form of (9), are the functions

$$y = (Ax + Bx^2)e^x. \qquad (10)$$

The derivatives of the functions (10) are

$$y = [Ax + Bx^2]e^x,$$
$$Dy = [A + (A + 2B)x + Bx^2]e^x,$$
$$D^2 y = [(2A + 2B) + (A + 4B)x + Bx^2]e^x,$$
$$D^3 y = [(3A + 6B) + (A + 6B)x + Bx^2]e^x.$$

Substitution in (9) gives

$$[(2A + 4B) + 4Bx + 0x^2]e^x = xe^x.$$

Therefore (10) is a solution of (9) if $4B = 1$ and $2A + 4B = 0$; that is, if $B = \frac{1}{4}$ and $A = -\frac{1}{2}$. The general solution of (9) is

$$y = (c_1 - \tfrac{1}{2}x + \tfrac{1}{4}x^2)e^x + c_2 \cos x + c_3 \sin x.$$

EXAMPLE 2.

$$(D^4 - D^3 + 4D^2 - 4D)y = D(D - 1)(D^2 + 4)y = x + \cos 2x. \qquad (11)$$

Since $D^2 x = 0$, and $(D^2 + 4) \cos 2x = 0$, we have

$$D^2(D^2 + 4)(x + \cos 2x) = 0.$$

Therefore all solutions of (11) are solutions of

$$D^3(D - 1)(D^2 + 4)^2 y = 0. \qquad (12)$$

The solutions of (12) which are not solutions of the reduced form of (11) are

$$y = Ax + Bx^2 + Cx \cos 2x + Ex \sin 2x. \tag{13}$$

Therefore, for some numbers A, B, C, E, determined by substitution in (11), (13) will be a particular solution of (11).

The following theorem can frequently be used to simplify the computations involved in the method of undetermined coefficients.

THEOREM 2. *If $L = L(D)$ is a linear differential operator (with constant coefficients) and F is any sufficiently differentiable function, then*

$$L(D)e^{rx}F(x) = e^{rx}L(D + r)F(x). \tag{14}$$

Proof. Let $L(D)$ be the operator

$$L(D) = p_n D^n + \cdots + p_1 D + p_0. \tag{15}$$

By $L(D + r)$ we mean the operator obtained by replacing D in (15) by $(D + r)$; that is,

$$L(D + r) = p_n(D + r)^n + \cdots + p_1(D + r) + p_0. \tag{16}$$

Let us first calculate $De^{rx}F(x)$, $D^2e^{rx}F(x)$, etc.

$$\begin{aligned} De^{rx}F(x) &= e^{rx}DF(x) + re^{rx}F(x) \\ &= e^{rx}(D + r)F(x). \end{aligned} \tag{17}$$

Using (17) again we get

$$\begin{aligned} D^2e^{rx}F(x) &= D[e^{rx}(D + r)F(x)] \\ &= e^{rx}(D + r)^2 F(x). \end{aligned} \tag{18}$$

Proceeding in this way, we find that

$$D^k e^{rx}F(x) = e^{rx}(D + r)^k F(x) \tag{19}$$

for all k. Therefore

$$\begin{aligned} L(D)e^{rx}F(x) &= [p_n D^n + \cdots + p_1 D + p_0]e^{rx}F(x) \\ &= p_n D^n e^{rx}F(x) + \cdots + p_1 De^{rx}F(x) + p_0 e^{rx}F(x) \\ &= p_n e^{rx}(D + r)^n F(x) + \cdots + p_1 e^{rx}(D + r)F(x) + p_0 e^{rx}F(x) \\ &= e^{rx}[p_n(D + r)^n + \cdots + p_1(D + r) + p_0]F(x) \\ &= e^{rx}L(D + r)F(x). \end{aligned}$$

COROLLARY. $L(D)e^{rx} = e^{rx}L(r)$.

Here $L(r)$ is the *number* obtained by substituting r for D in (15); that is,

$$L(r) = p_n r^n + \cdots + p_1 r + p_0.$$

The proof of the corollary is left to the student (Problem 11).

EXAMPLE 3. Let us use Theorem 2 to perform the calculations of Example 1. We know there is a solution of $(D - 1)(D^2 + 1)y = xe^x$ of the form $(Ax + Bx^2)e^x$. The substitution can be effected as follows:

$$
\begin{aligned}
(D - &1)(D^2 + 1)e^x(Ax + Bx^2) \\
&= e^x[(D + 1) - 1][(D + 1)^2 + 1](Ax + Bx^2) \\
&= e^x D(D^2 + 2D + 2)(Ax + Bx^2) \\
&= e^x(D^2 + 2D + 2)(A + 2Bx) \\
&= e^x(4B + 2A + 4Bx).
\end{aligned}
$$

EXAMPLE 4. Find a particular solution of $D^2(D - 1)(D^2 + 2)y = 3e^{2x}$. By Theorem 1 there is a solution of the form Ae^{2x}, and by the corollary above,

$$
\begin{aligned}
D^2(D - 1)(D^2 + 2)Ae^{2x} &= A D^2(D - 1)(D^2 + 2)e^{2x} \\
&= Ae^{2x}2^2(2 - 1)(2^2 + 2) = 24Ae^{2x}.
\end{aligned}
$$

Therefore $\frac{1}{8}e^{2x}$ is a particular solution.

PROBLEMS

Solve the following equations.

1. $(D^3 - D)y = 2e^x$
2. $(D^2 - 3D + 2)y = \cos x$
3. $(D^4 - 1)y = xe^x$
4. $(D^3 + D^2 - 2D)y = x^2$
5. $(D - 2)^2(D - 1)y = 4e^{2x}$
6. $(D^2 + 4)(D^2 + 1)y = 5 \sin x$
7. $(D^2 + 1)(D - 1)(D - 2)y = (x^2 + 1)e^x$
8. $(D - 1)(D^2 - 2D + 2)y = e^x \cos x$
9. $(D - 2)^2(D + 1)y = 18xe^{2x}$
10. Prove Theorem 1 for the case $b = 0$.
11. Prove the corollary to Theorem 2.

ANSWERS

1. $y = xe^x + c_1 + c_2e^x + c_3e^{-x}$
2. $y = \frac{1}{10}\cos x - \frac{3}{10}\sin x + c_1e^x + c_2e^{2x}$
3. $y = (-\frac{3}{8}x + \frac{1}{8}x^2)e^x + c_1e^x + c_2e^{-x} + c_3\cos x + c_4\sin x$

4. $y = -\frac{3}{4}x - \frac{1}{4}x^2 - \frac{1}{6}x^3 + c_1 + c_2 e^x + c_3 e^{-2x}$

5. $y = 2x^2 e^{2x} + c_1 e^x + (c_2 + c_3 x)e^{2x}$

6. $y = -\frac{5}{6}x \cos x + c_1 \cos x + c_2 \sin x + c_3 \cos 2x + c_4 \sin 2x$

7. $y = (-x - \frac{1}{6}x^3)e^x + c_1 \cos x + c_2 \sin x + c_3 e^x + c_4 e^{2x}$

8. $y = -\frac{1}{2}xe^x \cos x + c_1 e^x + c_2 e^x \cos x + c_3 e^x \sin x$

9. $y = (-x^2 + x^3)e^{2x} + (c_1 + c_2 x)e^{2x} + c_3 e^{-x}$

4–5 Inverse operators. There are many manipulations with operators which simplify the process of finding particular solutions for linear equations. We will look at a few techniques associated with the idea of an inverse operator. One inverse operator, the indefinite integral, is familiar from calculus. The notation $\int f(x)\,dx = F(x)$ is used to indicate that $D_x F(x) = f(x)$; in this sense, $\int(\ )\,dx$ is inverse to D_x. We extend this idea and describe an inverse for any linear differential operator $L(D)$.

AGREEMENT. We will write $y = [1/L(D)]q(x)$ or $y = L(D)^{-1}q(x)$ if $L(D)y = q(x)$. That is, we write $y = L(D)^{-1}q(x)$ to indicate that y is some particular solution of $L(D)y = q(x)$.

We define the sum and product of two inverse operators in the natural way,

$$[L_1(D)^{-1} + L_2(D)^{-1}]q(x) = L_1(D)^{-1}q(x) + L_2(D)^{-1}q(x), \quad (1)$$

$$[L_1(D)^{-1}L_2(D)^{-1}]q(x) = L_1(D)^{-1}[L_2(D)^{-1}q(x)]. \quad (2)$$

The following formulas, which express the linear properties of inverse operators, follow directly from Theorem 1 of Section 4–2.

$$L(D)^{-1}[q_1(x) + q_2(x)] = L(D)^{-1}q_1(x) + L(D)^{-1}q_2(x), \quad (3)$$

$$L(D)^{-1}[cq(x)] = cL(D)^{-1}q(x). \quad (4)$$

The commutative property (corollary to Theorem 3, Section 4–2) of linear differential operators, and the definition (2) above yield the formula

$$[L_1(D)L_2(D)]^{-1}q(x) = [L_2(D)^{-1}L_1(D)^{-1}]q(x)$$
$$= [L_1(D)^{-1}L_2(D)^{-1}]q(x). \quad (5)$$

Recall from Section 2–4 that the first order linear equation

$$(D - a)y = q(x) \quad (6)$$

has a particular solution

$$y = e^{ax}\int e^{-ax}q(x)\,dx. \quad (7)$$

This fact can be written

$$y = \frac{1}{D - a}\, q(x) = (D - a)^{-1}q(x) = e^{ax}\int e^{-ax}q(x)\, dx. \qquad (8)$$

From (8) and (5) it follows that a particular solution of

$$(D - a)(D - b)y = q(x) \qquad (9)$$

can be written

$$y = (D - a)^{-1}[(D - b)^{-1}q(x)]$$
$$= e^{ax}\int e^{-ax}\left[e^{bx}\int e^{-bx}q(x)\, dx\right] dx. \qquad (10)$$

The method indicated in (10) of expressing a particular solution in terms of repeated integrals can also be extended to equations of higher order.

EXAMPLE 1. $(D - 2)(D - 1)^2 y = e^x$.

A particular solution y can be obtained as follows:

$$y = (D - 2)^{-1}(D - 1)^{-1}[(D - 1)^{-1}e^x]$$
$$= (D - 2)^{-1}(D - 1)^{-1}[e^x\int e^{-x}e^x\, dx]$$
$$= (D - 2)^{-1}[(D - 1)^{-1}(xe^x)]$$
$$= (D - 2)^{-1}[e^x\int e^{-x}xe^x\, dx]$$
$$= (D - 2)^{-1}[\tfrac{1}{2}x^2 e^x]$$
$$= e^{2x}\int e^{-2x}\tfrac{1}{2}x^2 e^x\, dx$$
$$= \tfrac{1}{2}e^{2x}\int x^2 e^{-x}\, dx$$
$$= -\tfrac{1}{2}e^x(x^2 + 2x + 2).$$

If the inverse operators are applied in a different order, one gets a different particular solution; for example (Problem 1),

$$y = (D - 1)^{-1}(D - 2)^{-1}(D - 1)^{-1}e^x = -\tfrac{1}{2}e^x(x^2 + 2x).$$

These two solutions differ by $-\tfrac{1}{2}e^x$, which is a solution of the reduced equation $(D - 1)^2(D - 2)y = 0$.

Note that the process illustrated in the example really amounts to considering the system

$$(D - 1)u = e^x,$$
$$(D - 1)v = u,$$
$$(D - 2)y = v,$$

as we did in Examples 3 and 4 of Section 4–2, except here we ask for only one solution of each equation.

The method of Example 1 can be used to find a particular solution of any equation $L(D)y = q(x)$, even if some of the linear factors of $L(D)$ are complex. In most cases, however, the integrations involved become so involved that the method is not practical. The following theorem provides a method which usually leads to simpler calculations than (10).

THEOREM 1   (Partial fractions method).   *If a and b are distinct real numbers, then the following equation is an operator identity if it is a formal algebraic identity:*

$$\frac{1}{(D-a)(D-b)} = \frac{A}{D-a} + \frac{B}{D-b}. \tag{11}$$

Proof. The assumption is that A and B are the numbers such that $A(D-b) + B(D-a) = 1$. The assertion is that $Ay_1 + By_2$ is a solution of $(D-a)(D-b)y = q(x)$ if $y_1 = (D-a)^{-1}q(x)$ and $y_2 = (D-b)^{-1}q(x)$. Let us assume, therefore, that y_1 and y_2 are particular functions satisfying $(D-a)y_1 = q(x)$, and $(D-b)y_2 = q(x)$. Then

$$(D-a)(D-b)[Ay_1 + By_2]$$
$$= A(D-b)[(D-a)y_1] + B(D-a)[(D-b)y_2]$$
$$= A(D-b)q(x) + B(D-a)q(x)$$
$$= [A(D-b) + B(D-a)]q(x)$$
$$= [1]q(x) = q(x).$$

EXAMPLE 2. $(D+1)(D-2)y = xe^x$.

$$y = \frac{1}{(D+1)(D-2)} xe^x = \left[\frac{-\frac{1}{3}}{D+1} + \frac{\frac{1}{3}}{D-2}\right] xe^x$$

$$= -\frac{1}{3}\frac{1}{D+1} xe^x + \frac{1}{3}\frac{1}{D-2} xe^x$$

$$= -\frac{1}{3}e^{-x}\int e^x xe^x\, dx + \frac{1}{3}e^{2x}\int e^{-2x} xe^x\, dx$$

$$= -\frac{1}{3}e^{-x}[e^{2x}(\frac{1}{2}x - \frac{1}{4})] + \frac{1}{3}e^{2x}[e^{-x}(-x - 1)]$$

$$= -\frac{1}{12}e^x(2x - 1) - \frac{1}{12}e^x(4x + 4)$$

$$= -\frac{1}{4}e^x(2x + 1).$$

By way of review, let us check this last answer using Theorem 2 of Section 4–4.

$$(D+1)(D-2)[-\tfrac{1}{4}e^x(2x + 1)]$$
$$= -\tfrac{1}{4}e^x(D+2)(D-1)(2x + 1)$$
$$= -\tfrac{1}{4}e^x(D+2)(2 - 2x - 1)$$
$$= -\tfrac{1}{4}e^x(-2 - 4x + 2) = xe^x.$$

Theorem 1 extends to third and higher order operators. For example, if a, b, and c are distinct numbers, and A, B, and C are the numbers such that

$$\frac{1}{(D-a)(D-b)(D-c)} = \frac{A}{D-a} + \frac{B}{D-b} + \frac{C}{D-c} \tag{12}$$

is an algebraic identity, then

$$A(D-a)^{-1}q(x) + B(D-b)^{-1}q(x) + C(D-c)^{-1}q(x)$$

is a particular solution of $(D-a)(D-b)(D-c)y = q(x)$ (Problem 3). In the last sections we have derived the formulas

$$L(D)e^{rx}F(x) = e^{rx}L(D+r)F(x),$$
$$L(D)e^{rx} = e^{rx}L(r),$$
$$(D-r)^k x^k e^{rx} = k!e^{rx}.$$

The following formulas are simply restatements of those above in terms of inverse operators.

$$\frac{1}{L(D)} e^{rx}F(x) = e^{rx} \frac{1}{L(D+r)} F(x), \tag{13}$$

$$\frac{1}{L(D)} e^{rx} = \frac{e^{rx}}{L(r)}, \quad \text{(if } L(r) \neq 0\text{)}, \tag{14}$$

$$\frac{1}{(D-r)^k} e^{rx} = \frac{x^k e^{rx}}{k!}. \tag{15}$$

EXAMPLE 3. $(D-2)^3(D+2)^2(D-1)y = e^{2x}$.

A particular solution is given by

$$y = \frac{1}{(D-2)^3} \left[\frac{1}{(D+2)^2(D-1)} \right] e^{2x}$$

$$= \frac{1}{(D-2)^3} \left[\frac{e^{2x}}{16} \right] \qquad \text{[from (14)]}$$

$$= \frac{1}{16} \frac{x^3 e^{2x}}{3!} = \frac{1}{96} x^3 e^{2x} \qquad \text{[from (15)].}$$

PROBLEMS

1. Find a solution of the equation of Example 1 by evaluating the inverse operators in this order: $(D-1)^{-1}(D-2)^{-1}(D-1)^{-1}e^x$.

2. Use (10) to find a solution of $(D+1)^2y = e^{-x}\ln|x|$ and check your answer.

3. Show that (12) is an operator identity if it is an algebraic identity (compare the proof of Theorem 1).

4. Use the partial fractions method (12) to solve

$$(D + 1)(D - 1)(D - 2)y = e^x.$$

5. Find a particular solution of $(D - 1)(D + 2)y = e^{-2x}$ and use (14) of Section 4-4 to check your answer.

(a) use formula (10)   (b) use formula (13)

6. Use (14) and (15) to solve the following:

(a) $(D - 2)^2(D + 1)y = e^{2x}$ (b) $D(D + 1)(D - 2)^2y = e^{-x}.$

7. Let r be a complex number and $\bar{r}$ its conjugate. Show that formula (10) can be used to find the real solutions

$$y = \frac{1}{r\bar{r}}\, x + \frac{1}{(r\bar{r})^2}\,(r + \bar{r}) \qquad \text{of} \qquad (D - r)(D - \bar{r})y = x.$$

8. Use the formula of Problem 7 to solve $(D^2 - 2D + 5)y = x.$

9. Use Theorem 1 to find a particular solution of $(D^2 - 3D + 2)y = \sin^2 x.$

$$\left[\textit{Hint:} \int e^{ax} \sin^2 x \, dx = \frac{e^{ax}}{a^2 + 4}\left[a \sin^2 x - 2 \sin x \cos x + \frac{2}{a}\right]. \right]$$

Answers

1. $y = -\frac{1}{2}e^x(x^2 + 2x)$
2. $y = \frac{1}{4}x^2e^{-x}(2\ln|x| - 3)$
4. $y = -\frac{1}{4}e^x(1 + 2x)$
5. (a) and (b) $y = -\frac{1}{9}e^{-2x}(3x + 1)$
6. (a) $y = \frac{1}{6}x^2e^{2x}$ (b) $y = -\frac{1}{9}xe^{-x}$
8. $y = \frac{1}{5}x + \frac{2}{25}$
9. $y = -\frac{1}{20}\sin^2 x + \frac{3}{20}\sin x \cos x + \frac{11}{40}$

4-6 Variation of parameters method. In this section we give a method for finding a particular solution of the linear equation

$$y^{(n)} + \cdots + p_1 y' + p_0 y = q \tag{1}$$

whenever we have n linearly independent solutions of the reduced equation. Here we are *not* assuming that the coefficients $p_0, p_1, \ldots, p_{n-1}$ are constants, but that these are arbitrary functions continuous on some common interval. Since our concern so far has been primarily with equations with constant coefficients, let us recall that the theory of Section 3-3 applies to any linear equation, whether the coefficients are constants or functions.

Thus to solve (1), we must find n linearly independent solutions of the reduced equation and one solution of (1). Although we give no general methods for solving the reduced equation if the coefficients are not constant, the variation of parameters method insures that *if* we can solve the reduced equation, then we can solve (1) for any right-hand member q.

Let us first illustrate the method for the second order equation.

$$y'' + p_1 y' + p_0 y = q. \tag{2}$$

The assumption is that we have two linearly independent solutions y_1, y_2 of the reduced equation

$$y'' + p_1 y' + p_0 y = 0. \tag{3}$$

We try to find a solution y of (2) in the form

$$y = v_1 y_1 + v_2 y_2, \tag{4}$$

where v_1 and v_2 are functions to be determined. In general, there are many functions v_1, v_2 such that $v_1 y_1 + v_2 y_2$ is a solution of (2) (see Problem 1). Since it is reasonable to assume that we can put additional conditions on v_1 and v_2, we make an assumption which facilitates the computations. We show that there are functions v_1 and v_2 such that the function $y = v_1 y_1 + v_2 y_2$ is a solution of (2) and also satisfies the equation

$$y' = v_1 y_1' + v_2 y_2'. \tag{5}$$

This last assumption is clearly equivalent to the condition

$$y_1 v_1' + y_2 v_2' = 0.$$

Computing y'' from (5), we get

$$y'' = v_1 y_1'' + v_2 y_2'' + y_1' v_1' + y_2' v_2'. \tag{6}$$

Now we substitute y, y', and y'' as given by (4), (5), and (6) in Eq. (2) and group the terms containing v_1 and those containing v_2. The result is

$$v_1[y_1'' + p_1 y_1' + p_0 y_1] + v_2[y_2'' + p_1 y_2' + p_0 y_2] + y_1' v_1' + y_2' v_2' = q. \tag{7}$$

The terms in brackets are zero, since y_1 and y_2 are solutions of (3), and the condition that y be a solution of (2) is therefore

$$y_1' v_1' + y_2' v_2' = q.$$

That is, the function y of (4) will have the derivative (5) and be a solution of (2), if v_1 and v_2 satisfy the equations

$$y_1 v_1' + y_2 v_2' = 0,$$
$$y_1' v_1' + y_2' v_2' = q. \tag{8}$$

For each x, the equations (8) are simultaneous linear equations in $v_1'(x)$ and $v_2'(x)$. The determinant of this system is the Wronskian of the two known functions y_1 and y_2:

$$W(y_1(x), y_2(x)) = \begin{vmatrix} y_1(x) & y_2(x) \\ y_1'(x) & y_2'(x) \end{vmatrix}.$$

Since y_1 and y_2 are linearly independent solutions of the reduced equation, the Wronskian is never zero, and the equations (8) have the solutions

$$v_1'(x) = -y_2(x)q(x)/W(y_1(x), y_2(x)),$$
$$v_2'(x) = y_1(x)q(x)/W(y_1(x), y_2(x)). \tag{9}$$

Since the functions on the right of equations (9) are continuous, they have antiderivatives, and there are functions v_1 and v_2 satisfying (9) and hence (8). For these functions, $v_1 y_1 + v_2 y_2$ is a particular solution of (2), and the general solution of (2) is

$$y = (c_1 + v_1)y_1 + (c_2 + v_2)y_2.$$

EXAMPLE 1. $y'' - y = e^{2x}$.

Two linearly independent solutions of $y'' - y = 0$ are e^x and e^{-x}. Let $y = v_1 e^x + v_2 e^{-x}$, and suppose that $y' = v_1 e^x - v_2 e^{-x}$; that is, suppose that $e^x v_1' + e^{-x} v_2' = 0$. Then

$$y'' = v_1 e^x + v_2 e^{-x} + e^x v_1' - e^{-x} v_2'.$$

Substitution in the differential equation gives

$$v_1[e^x - e^x] + v_2[e^{-x} - e^{-x}] + e^x v_1' - e^{-x} v_2' = e^{2x}.$$

In other words, y will have the given derivative, and be a solution, if

$$e^x v_1' + e^{-x} v_2' = 0,$$

and

$$e^x v_1' - e^{-x} v_2' = e^{2x}.$$

Solving these equations, we get $v_1' = \tfrac{1}{2}e^x$, $v_1 = \tfrac{1}{2}e^x$, and $v_2' = -\tfrac{1}{2}e^{3x}$, $v_2 = -\tfrac{1}{6}e^{3x}$. The equation therefore has a particular solution

$$y = \tfrac{1}{2}e^x e^x - \tfrac{1}{6}e^{3x}e^{-x} = \tfrac{1}{3}e^{2x}.$$

EXAMPLE 2. $y'' + y = \sec x$.

Two solutions of the reduced equation are $\cos x$ and $\sin x$. We set

$$y = v_1 \cos x + v_2 \sin x,$$
$$y' = -v_1 \sin x + v_2 \cos x,$$
$$y'' = -v_1 \cos x - v_2 \sin x - v_1' \sin x + v_2' \cos x.$$

Then y will be a solution with the derivatives given above if

$$v_1' \cos x + v_2' \sin x = 0,$$
$$-v_1' \sin x + v_2' \cos x = \sec x.$$

The determinant of this system is $W(\cos x, \sin x) = 1$, and solving, we get

$$v_1' = \begin{vmatrix} 0 & \sin x \\ \sec x & \cos x \end{vmatrix} = -\frac{\sin x}{\cos x},$$

$$v_2' = \begin{vmatrix} \cos x & 0 \\ -\sin x & \sec x \end{vmatrix} = 1.$$

Therefore $v_1 = \ln|\cos x|$, $v_2 = x$, and a particular solution of the equation is

$$y = \cos x \ln|\cos x| + x \sin x.$$

Now we will generalize these ideas so that they apply to a linear equation of any order.

THEOREM 1. *If $y_1, \ldots, y_n$ are linearly independent solutions of the reduced form of*

$$y^{(n)} + p_{n-1} y^{(n-1)} + \cdots + p_1 y' + p_0 y = q, \tag{10}$$

then there are functions $v_1, \ldots, v_n$ which satisfy

$$
\begin{aligned}
y_1 v_1' \quad &+ \cdots + y_n v_n' &= 0, \\
y_1' v_1' \quad &+ \cdots + y_n' v_n' &= 0, \\
&\;\;\vdots & \tag{11} \\
y_1^{(n-2)} v_1' &+ \cdots + y_n^{(n-2)} v_n' &= 0, \\
y_1^{(n-1)} v_1' &+ \cdots + y_n^{(n-1)} v_n' &= q.
\end{aligned}
$$

and for any such functions $v_1, \ldots, v_n$, the function $y = v_1 y_1 + \cdots + v_n y_n$ is a solution of (10).

Proof. The system (11) of linear equations can be solved for $v_1', \ldots, v_n'$, since the determinant of the system is $W(y_1, \ldots, y_n)$, which never vanishes. Moreover, the solutions $v_1', \ldots, v_n'$ can be written as quotients of determinants whose entries are continuous functions. Hence the expressions for $v_1', \ldots, v_n'$ will be continuous, and will have antiderivatives $v_1, \ldots, v_n$. Therefore, there are functions satisfying (11).

The first $(n-1)$ of the equations (11) are the conditions that the first $(n-1)$ derivatives of y are those given below; the nth derivative of y is simply calculated from $y^{(n-1)}$ without further assumptions.

$$
\begin{aligned}
y &= v_1 y_1 & + \cdots + v_n y_n, \\
y' &= v_1 y_1' & + \cdots + v_n y_n', \\
y^{(n-1)} &= v_1 y_1^{(n-1)} + \cdots + v_n y_n^{(n-1)}, \\
y^{(n)} &= v_1 y_1^{(n)} & + \cdots + v_n y_n^{(n)} + y_1^{(n-1)} v_1' + \cdots + y_n^{(n-1)} v_n'.
\end{aligned}
\tag{12}
$$

Now substitute $y, y', \ldots, y^{(n)}$ from (12) into equation (10). The coefficient of v_1 after substitution is the first column of the array (12) with the appropriate coefficients; namely,

$$
y_1^{(n)} + p_{n-1} y_1^{(n-1)} + \cdots + p_1 y_1' + p_0 y_1.
\tag{13}
$$

Since y_1 is a solution of the reduced form of (10), the coefficient (13) of v_1 is zero. Similarly, the terms containing $v_2, \ldots, v_n$ drop out, since $y_2, \ldots, y_n$ are solutions of the reduced equation. The condition that y be a solution of (10) is therefore simply

$$
y_1^{(n-1)} v_1' + \cdots + y_n^{(n-1)} v_n' = q.
$$

This equation is satisfied by the last of the assumptions (11), and hence y is a solution of (10).

EXAMPLE 3. $y''' - 3y'' + 2y' = e^{-x}$.

The functions 1, e^x, e^{2x} are linearly independent solutions of the reduced equation. We find the functions v_1, v_2, and v_3 such that if

$$
y = v_1 + v_2 e^x + v_3 e^{2x},
$$

then

$$
\begin{aligned}
y' &= v_2 e^x + 2v_3 e^{2x} && \text{(that is, } v_1' + e^x v_2' + e^{2x} v_3' = 0\text{)}, \\
y'' &= v_2 e^x + 4v_3 e^{2x} && \text{(that is, } e^x v_2' + 2e^{2x} v_3' = 0\text{)}, \\
y''' &= v_2 e^x + 8v_3 e^{2x} + e^x v_2' + 4e^{2x} v_3',
\end{aligned}
$$

and

$$
y \text{ is a solution} \qquad \text{(that is, } e^x v_2' + 4e^{2x} v_3' = e^{-x}\text{)}.
$$

The determinant of the system of equations on the right above [equations (11) for this example] is

$$W(1, e^x, e^{2x}) = \begin{vmatrix} 1 & e^x & e^{2x} \\ 0 & e^x & 2e^{2x} \\ 0 & e^x & 4e^{2x} \end{vmatrix} = 2e^{3x}.$$

Hence,

$$v_1' = \frac{1}{2e^{3x}} \begin{vmatrix} 0 & e^x & e^{2x} \\ 0 & e^x & 2e^{2x} \\ e^{-x} & e^x & 4e^{2x} \end{vmatrix} = \tfrac{1}{2}e^{-x}.$$

Similarly, $v_2' = -2e^x/(2e^{3x}) = -e^{-2x}$, and $v_3' = 1/(2e^{3x}) = \tfrac{1}{2}e^{-3x}$. Integration gives $v_1 = -\tfrac{1}{2}e^{-x}$, $v_2 = \tfrac{1}{2}e^{-2x}$, and $v_3 = -\tfrac{1}{6}e^{-3x}$. A particular solution of the equation is therefore

$$y = -\tfrac{1}{2}e^{-x} + \tfrac{1}{2}e^{-2x}e^x - \tfrac{1}{6}e^{-3x}e^{2x} = -\tfrac{1}{6}e^{-x}.$$

Problems

1. Find functions v_1 and v_2 such that $v_1(x)e^x + v_2(x)e^{2x}$ is a solution of $y'' - 3y' + 2y = 4x - 4$, and such that

(a) $v_1(x) = 0$ (b) $v_1(x)y_1(x) = v_2(x)y_2(x)$

(c) v_1 and v_2 satisfy (8).

2. Write out in detail the statement and proof of Theorem 1 for the case $n = 3$.

3. Solve by variation of parameters: $y'' - y = xe^x$.

4. Solve the equation $y'' + y = \csc x$.

5. Check that e^x is a solution of $(1 - x)y'' + xy' - y = 0$. Find a second (polynomial) solution of the reduced equation and solve $(1 - x)y'' + xy' - y = (1 - x)^2$. [*Caution:* Do the equations (9) and (11) apply directly to the given equation in its present form?]

6. Show that x and $1/x$ are solutions of $y'' + (1/x)y' - (1/x^2)y = 0$. Solve by variation of parameters: $y'' + (1/x)y' - (1/x^2)y = \ln x$ $(x > 0)$.

7. Solve by variation of parameters: $y''' - y'' = x^3$.

8. Solve by variation of parameters: $y''' - 2y'' - y' + 2y = e^x$.

9. If y_0 is one solution of the reduced form of a linear equation, then the substitution $y = y_0u$ gives a linear equation in u in which u itself is missing. Thus the order is effectively reduced by one (substitute $v = u'$). Illustrate this for the second order equation $y'' + p_1y' + p_0y = q$.

10. Solve by the method of Problem 9.

(a) Problem 5, using $y_0 = e^x$;   (b) Problem 6, using $y_0 = x$.

11. Use the substitution of Problem 9 to solve $(x^2 - x)y' + (1 - 2x)y = -x^2$, after checking that $y_0 = x^2 - x$ is a solution of the reduced form.

ANSWERS

1. (a) $v_2(x) = (2x + 1)e^{-2x}$
    (b) $v_1(x) = (x + \frac{1}{2})e^{-x}$
    (c) $v_1(x) = 4xe^{-x}$, $v_2(x) = -(2x - 1)e^{-2x}$
3. $y = c_1e^x + c_2e^{-x} + \frac{1}{8}e^x(2x^2 - 2x + 1)$
4. $y = (c_1 - x)\cos x + (c_2 + \ln|\sin x|)\sin x$
5. $y = c_1e^x + c_2x + 1 + x + x^2$
6. $v_1 = \frac{1}{2}x(\ln x - 1)$, $v_2 = -\frac{1}{6}x^3 \ln x + \frac{1}{18}x^3$,
    $y = c_1x + c_2/x + \frac{1}{3}x^2 \ln x - \frac{4}{9}x^2$
7. $v_1 = \frac{1}{5}x^5 - \frac{1}{4}x^4$, $v_2 = -\frac{1}{4}x^4$, $v_3 = e^{-x}(-x^3 - 3x^2 - 6x - 6)$,
    $y = -\frac{1}{20}x^5 - \frac{1}{4}x^4 - x^3 - 3x^2 - 6x - 6 + c_1 + c_2x + c_3e^x$
8. $v_1 = -\frac{1}{2}x$, $v_2 = \frac{1}{12}e^{2x}$, $v_3 = -\frac{1}{3}e^{-x}$,
    $y = c_1e^x + c_2e^{-x} + c_3e^{2x} - \frac{1}{4}e^x(2x + 1)$
11. $y = x + c(x^2 - x)$

CHAPTER 5

THE LAPLACE TRANSFORM

5-1 Review of improper integrals. The Laplace transform is defined in terms of an improper integral, and we will review here some of the facts we will need about such integrals.

Recall that the Riemann integral

$$\int_a^b f(x)\, dx$$

is initially defined only for finite intervals $[a, b]$, and only for functions f which are bounded on the given interval. This definition is then extended to the so-called improper integrals, in which the integrand is unbounded on some finite interval, or the interval of integration is infinite. We will be concerned only with integrals which are improper because the upper limit of integration is infinite. Specifically, we consider integrals of the form

$$\int_0^\infty f(x)\, dx, \tag{1}$$

where f is continuous on $[0, \infty)$. The integral (1) is defined as the limit

$$\int_0^\infty f(x)\, dx = \lim_{R \to \infty} \int_0^R f(x)\, dx. \tag{2}$$

If this limit exists, we say the integral (1) *converges*, otherwise we say the integral *diverges*. The following comparison test, similar to that for series, is a basic method for showing the convergence of (1).

THEOREM 1. *If $|f(x)| \leq g(x)$ for all sufficiently large x, and $\int_0^\infty g(x)\, dx$ converges, then $\int_0^\infty f(x)\, dx$ converges.*

Notice that, in particular, Theorem 1 says that $\int_0^\infty f(x)\, dx$ converges if $\int_0^\infty |f(x)|\, dx$ converges.

Now consider the case in which the integrand contains a parameter. Suppose $f(s, x)$ is continuous for $\alpha \leq s \leq \beta$ and $x \geq 0$ and that the following integral converges for each s in $[\alpha, \beta]$.

$$\int_0^\infty f(s, x)\, dx = \phi(s). \tag{3}$$

If we write

$$F_R(s) = \int_0^R f(s, x) \, dx, \tag{4}$$

then (3) is the same as saying that for each s in $[\alpha, \beta]$

$$\lim_{R \to \infty} F_R(s) = \phi(s). \tag{5}$$

That is, for each s in $[\alpha, \beta]$ and every $\epsilon > 0$ there is a number N (depending on s and ϵ) such that

$$|F_R(s) - \phi(s)| < \epsilon \tag{6}$$

whenever $R \geq N$. The situation in (5), or (3), is similar to the convergence of a sequence of functions $\{F_n\}$, except here we have a function F_R for each number $R \geq 0$.

The function ϕ defined in (3) need not be continuous unless some assumption is made about how the convergence of the integral depends on s. We say the integral (3) *converges uniformly for s in $[\alpha, \beta]$* if for every $\epsilon > 0$ there is a number N (depending only on ϵ) such that $|F_R(s) - \phi(s)| < \epsilon$ for all s in $[\alpha, \beta]$ if $R \geq N$.

THEOREM 2. *If the integral* (3) *converges uniformly for s in $[\alpha, \beta]$, then ϕ is continuous on $[\alpha, \beta]$.*

The following theorem provides a convenient test for uniform convergence.

THEOREM 3 (Weierstrass M-test for integrals). *If M is a continuous function on $[0, \infty)$ such that $\int_0^\infty M(x) \, dx$ converges, and $|f(s, x)| \leq M(x)$ for all s in $[\alpha, \beta]$ and all $x \geq 0$, then $\int_0^\infty f(s, x) \, dx$ converges uniformly on $[\alpha, \beta]$.*

Uniform convergence is also the critical hypothesis for showing that the function ϕ of (3) is differentiable, and that $\phi'(s)$ can be found by differentiating under the integral sign in (3).

THEOREM 4. *If $f_s(s, x)$ is continuous for s in $[\alpha, \beta]$ and $x \geq 0$, and the integral*

$$\int_0^\infty f_s(s, x) \, dx.$$

converges uniformly on $[\alpha, \beta]$, then ϕ is differentiable and

$$\phi'(s) = \int_0^\infty f_s(s, x) \, dx.$$

Note that it is the integral of the partial derivative $f_s(s, x)$ which must converge uniformly.

EXAMPLE 1. Find the values of s for which the integral

$$\phi(s) = \int_0^\infty (1 + s^2)e^{-sx} \, dx \qquad (7)$$

converges, and find a formula for $\phi(s)$.

For $s = 0$, the integrand is identically one, and the integral diverges. If $s \neq 0$, then

$$\int_0^\infty (1 + s^2)e^{-sx} \, dx = \lim_{R \to \infty} (1 + s^2)\int_0^R e^{-sx} \, dx$$

$$= \lim_{R \to \infty} -\frac{(1 + s^2)}{s} (e^{-sR} - 1).$$

This last limit exists if $s > 0$ ($\lim_{R \to \infty} e^{-sR} = 0$), and fails to exist if $s < 0$ ($\lim_{R \to \infty} e^{-sR} = \infty$). Therefore the integral (7) converges if and only if $s > 0$, and for these values of s

$$\phi(s) = \frac{(1 + s^2)}{s}.$$

EXAMPLE 2. Show that the function ϕ defined by

$$\phi(s) = \int_0^\infty e^{-sx} \, dx \qquad (8)$$

is defined and differentiable for $s > 0$.

As in the example above it is easy to show that the integral in (8) converges if and only if $s > 0$. Since $(\partial/\partial s)e^{-sx} = -xe^{-sx}$, we want to show that

$$\phi'(s) = \int_0^\infty -xe^{-sx} \, dx \qquad (9)$$

for any $s > 0$. To verify (9), we show that the integral of (9) converges uniformly on some interval $[\alpha, \beta]$ around s, for each $s > 0$. For any numbers α and β with $0 < \alpha < \beta$ and any number s in $[\alpha, \beta]$, we have

$$|-xe^{-sx}| = xe^{-sx} \leq xe^{-\alpha x},$$

and the integral

$$\int_0^\infty xe^{-\alpha x} \, dx \qquad (10)$$

converges (Problem 1). By Theorem 3, with $M(x) = xe^{-\alpha x}$, the integral in (9) converges uniformly on $[\alpha, \beta]$. It then follows from Theorem 4 that $\phi'(s)$ exists and is given by (9) for all s in $[\alpha, \beta]$. Since any number $s > 0$ is in some interval $[\alpha, \beta]$ with $\alpha > 0$, $\phi'(s)$ exists for all $s > 0$.

PROBLEMS

1. Show that (10) converges, given that $\alpha > 0$.

2. Evaluate the integrals in (8) and (9), and verify that $\phi'(s)$ is given by (9).

3. Let $\phi(s) = \int_0^\infty x e^{-sx}\, dx$. Show that $\phi(s)$ is defined for $s > 0$ and that $\phi'(s)$ exists for $s > 0$.

4. Let p be a fixed number with $0 < p < 1$. Verify that the following integral converges uniformly for s in $[p, 1]$:

$$\int_0^\infty \frac{\sin (sx)}{s}\, e^{-x}\, dx.$$

5. Show that the integral of Problem 4 can be differentiated under the integral sign for s in $[p, 1]$.

5-2 The Laplace transform. Suppose that f is a function on $[0, \infty)$ such that the following integral converges for some values of s.

$$\int_0^\infty e^{-sx} f(x)\, dx. \tag{1}$$

The integral (1) depends on the function f and on the number s. We regard (1) as determining a new function $\hat{f}$ associated with f, where $\hat{f}$ is defined by

$$\hat{f}(s) = \int_0^\infty e^{-sx} f(x)\, dx \tag{2}$$

for those values of s for which the integral converges. The function $\hat{f}$ is called the Laplace transform of f. We will also use the notation

$$\mathcal{L}(f(x)) = \hat{f}(s)$$

to indicate the relationship (2). Thus $\mathcal{L}$ is an operator, defined on a class of functions f that we will call *object functions*, with values $\hat{f}$ which are functions that we will refer to as *transforms*.

The following three theorems give properties of $\mathcal{L}$ which are of primary importance in the application of Laplace transforms to linear differential equations.

THEOREM 1. $\mathcal{L}$ *is a linear operator; i.e., for any s such that $\hat{f}(s)$ and $\hat{g}(s)$ are both defined, and any numbers a and b,*

$$\mathcal{L}(af(x) + bg(x)) = a\mathcal{L}(f(x)) + b\mathcal{L}(g(x)) = a\hat{f}(s) + b\hat{g}(s).$$

Proof. This theorem is just a restatement of the linear properties of the defining integral (2).

The following theorem will be proved in the next section.

THEOREM 2. *If f and g are continuous on* $[0, \infty)$ *and* $\hat{f}(s) \equiv \hat{g}(s)$, *then* $f(x) \equiv g(x)$.

The point of Theorem 2 is that the operator $\mathcal{L}$ sets up a *one-to-one* correspondence between continuous object functions f and their transforms $\hat{f}$. In other words, if we can determine what $\hat{f}$ is, then we know what f must be.

THEOREM 3. *If* f' *is continuous on* $[0, \infty)$ *and* $\lim_{x \to \infty} e^{-sx}f(x) = 0$, *then* $\mathcal{L}(f'(x))$ *exists at s if and only if* $\mathcal{L}(f(x))$ *exists at s, and*

$$\mathcal{L}(f'(x)) = s\mathcal{L}(f(x)) - f(0) = s\hat{f}(s) - f(0). \tag{3}$$

Proof. Let s be any number such that $\lim_{x \to \infty} e^{-sx}f(x) = 0$. Integration by parts gives the following formula for any number $R > 0$:

$$\int_0^R e^{-sx}f'(x)\, dx = e^{-sx}f(x)\Big]_{x=0}^R + \int_0^R se^{-sx}f(x)\, dx$$

$$= e^{-sR}f(R) - f(0) + s\int_0^R e^{-sx}f(x)\, dx.$$

Since $e^{-sR}f(R) \to 0$ as $R \to \infty$ by hypothesis, we have

$$\lim_{R \to \infty} \int_0^R e^{-sx}f'(x)\, dx = -f(0) + \lim_{R \to \infty} s\int_0^R e^{-sx}f(x)\, dx.$$

That is, if either limit above exists, so does the other, and the given equality holds. This says that $\mathcal{L}(f'(x))$ exists at s if and only if $\mathcal{L}(f(x))$ does, and if either transform exists, then (3) holds.

COROLLARY. *Assume f is continuous on* $[0, \infty)$ *and* $F(x) = \int_0^x f(t)\, dt$. *If* $\hat{f}(s)$ *exists and* $\lim_{x \to \infty} e^{-sx}F(x) = 0$, *then* $\hat{F}(s)$ *exists and* $\hat{F}(s) = (1/s)\hat{f}(s)$, *that is,*

$$\mathcal{L}\left(\int_0^x f(t)\, dt\right) = \frac{1}{s}\,\mathcal{L}(f(x)). \tag{4}$$

Proof. This follows from Theorem 3, since $F'(x) = f(x)$ and $F(0) = 0$.

Note that differentiation of an object function corresponds to an algebraic operation (3) on its transform. We show below how this principle is used to solve a linear differential equation with constant coefficients. First we compute the transforms of some simple functions in the following examples.

EXAMPLE 1.

$$\mathcal{L}(1) = \frac{1}{s} \quad \text{for } s > 0.$$

Directly from the definition and the assumption that $s > 0$, we have

$$\mathcal{L}(1) = \int_0^\infty e^{-sx}1 \, dx = \lim_{R\to\infty} -\frac{1}{s}(e^{-sR} - 1) = \frac{1}{s}\cdot$$

It is clear that the integral above diverges for $s \leq 0$, so $\mathcal{L}(1)$ is defined only for $s > 0$.

EXAMPLE 2.

$$\mathcal{L}(x) = \frac{1}{s^2} \quad \text{and} \quad \mathcal{L}(x^2) = \frac{2}{s^3} \quad \text{for } s > 0.$$

These formulas can be found directly from the definition as in the example above, but we will instead derive them using the corollary of Theorem 3. Let $f(x) = 1$, so that $\int_0^x f(t) \, dt = x = F(x)$. For every $s > 0$, $\lim_{x\to\infty} e^{-sx}F(x) = \lim_{x\to\infty} xe^{-sx} = 0$, so

$$\mathcal{L}(F(x)) = \frac{1}{s}\mathcal{L}(f(x));$$

that is,

$$\mathcal{L}(x) = \frac{1}{s}\mathcal{L}(1) = \frac{1}{s^2}\cdot$$

Similarly, if $f(x) = 2x$, then $F(x) = x^2$, and again

$$\lim_{x\to\infty} e^{-sx}F(x) = \lim_{x\to\infty} x^2 e^{-sx} = 0 \quad \text{if } s > 0.$$

Hence

$$\mathcal{L}(x^2) = \frac{1}{s}\mathcal{L}(2x) = \frac{2}{s}\mathcal{L}(x) = \frac{2}{s}\frac{1}{s^2} = \frac{2}{s^3}\cdot$$

The transforms of x^n, $n = 1, 2, 3, \ldots$, can be calculated by repeating this process (Problem 2).

EXAMPLE 3.

$$\mathcal{L}(e^x) = \frac{1}{s - 1} \quad \text{for } s > 1.$$

Assume that $s > 1$, so that $e^{(1-s)R} \to 0$ as $R \to \infty$. Then we have

$$\mathcal{L}(e^x) = \int_0^\infty e^{-sx} e^x \, dx = \int_0^\infty e^{(1-s)x} \, dx$$

$$= \lim_{R \to \infty} \frac{1}{1-s} \left(e^{(1-s)R} - 1 \right) = \frac{1}{s-1}.$$

Now we can illustrate how the Laplace transform is used to solve a linear differential equation. Consider the equation

$$y' - y = 1 - x, \qquad y(0) = 2. \tag{5}$$

Let y be the solution of (5), and assume* that $\mathcal{L}(y(x))$ exists and $\lim_{x \to \infty} e^{-sx} y(x) = 0$ for all sufficiently large values of s. Then $\mathcal{L}(y'(x))$ exists, and

$$\mathcal{L}(y'(x) - y(x)) = \mathcal{L}(1 - x),$$

$$\mathcal{L}(y'(x)) - \mathcal{L}(y(x)) = \mathcal{L}(1) - \mathcal{L}(x),$$

$$s\mathcal{L}(y(x)) - y(0) - \mathcal{L}(y(x)) = \frac{1}{s} - \frac{1}{s^2},$$

$$\mathcal{L}(y(x))(s - 1) = \frac{1}{s} - \frac{1}{s^2} + 2,$$

$$\mathcal{L}(y(x)) = \frac{s - 1 + 2s^2}{s^2(s - 1)} = \frac{1}{s^2} + \frac{2}{s - 1}.$$

Since

$$\mathcal{L}(x) = 1/s^2, \text{ and } \mathcal{L}(e^x) = 1/(s - 1),$$

we have

$$\mathcal{L}(y(x)) = \mathcal{L}(x + 2e^x).$$

From Theorem 2 it follows that

$$y(x) = x + 2e^x.$$

The use of the transform to solve a linear equation, as illustrated above, can be thought of in the following way. Imagine (Theorem 2) a table in which we can find $y(x)$ given $\hat{y}(s)$. By Theorems 1 and 3, a constant co-

---

* We will show in the next section that these assumptions are always satisfied for a solution of a linear equation with constant coefficients.

efficient linear differential equation in $y(x)$ is transformed into an algebraic equation in $\hat{y}(s)$, involving the given initial condition. We solve the algebraic equation to find $\hat{y}(s)$, consult the table to find $y(x)$, and the problem is solved. Note that the particular solution for the given initial condition is found without first finding the general solution. If we leave the initial condition arbitrary, $y(0) = c$, then we obtain the general solution $y = x + ce^x$.

PROBLEMS

1. Show by induction that if $s > 0$, then $\lim_{x \to \infty} x^n e^{-sx} = 0$ for all positive integers n. [*Hint:* Write $x^{n+1} e^{-sx} = x^{n+1}/e^{sx}$ and use l'Hospital's rule.]

2. Use Problem 1 and Theorem 3 to prove that $\mathcal{L}(x^n) = n!/s^{n+1}$ for $s > 0$ and $n = 0, 1, 2, \ldots$.

3. (a) Show that $\mathcal{L}(e^{ax}) = 1/(s - a)$ if $s > a$.

   (b) Find $\mathcal{L}(\cosh ax)$ and $\mathcal{L}(\sinh ax)$ from part (a) and Theorem 1.

4. Show directly from the definition (2) that

$$\mathcal{L}(\sin(ax)) = a/(a^2 + s^2) \quad \text{if } s > 0.$$

5. Use Problem 4 and Theorem 3 to find $\mathcal{L}(\cos ax)$.

6. Let $f(x) = xe^{ax}$, so that $f'(x) = af(x) + e^{ax}$, and $\mathcal{L}(f'(x)) = a\mathcal{L}(f(x)) + 1/(s - a)$. Use this equation and Theorem 3 to find $\mathcal{L}(xe^{ax})$ without integration.

7. (a) Find $\mathcal{L}(x^2 e^{ax})$ by the technique of Problem 6.

   (b) Show by induction that $\mathcal{L}(x^n e^{ax}) = n!/(s + a)^{n+1}$ for $s > a$, and $n = 1, 2, \ldots$.

8. Solve the equation $y' + y = 1$, $y(0) = 2$, by showing that

$$y(s) = \frac{1 + 2s}{s(s + 1)} = \frac{1}{s} + \frac{1}{s + 1}.$$

[cf. Problem 3(a).]

9. Use the transform to solve $y' - 2y = 1 - 2x$, $y(0) = 1$.

10. Show that with suitable hypotheses,

$$\mathcal{L}(y''(x)) = s^2 \mathcal{L}(y(x)) - sy(0) - y'(0).$$

11. Find $\hat{y}(s)$ if y is the solution of $y'' + 2y' + y = e^x + 1$, $y(0) = 1$, $y'(0) = 0$.

ANSWERS

3. (b) $\mathcal{L}(\cosh ax) = s/(s^2 - a^2)$, $\mathcal{L}(\sinh ax) = a/(s^2 - a^2)$

5. $\mathcal{L}(\cos ax) = s/(s^2 + a^2)$ 6. $\mathcal{L}(xe^{ax}) = 1/(s - a)^2$

7. $\mathcal{L}(x^2 e^{ax}) = 2/(s - a)^3$ 8. $y = 1 + e^{-x}$

9. $y = x + e^{2x}$

11. $\hat{y}(s) = (s^3 + s^2 - 1)/s(s - 1)(s + 1)^2$

5–3 Properties of the transform. The discussion of the preceding section is largely formal, and in this section we prove the theorems which allow us to deal with transforms in a systematic way.

We will assume from the beginning that all object functions are continuous on $[0, \infty)$. This is not a necessary restriction, but it will simplify the discussion. A continuous function will have a transform defined for some values of s provided the function does not grow too rapidly as $x \to \infty$. We say that f is of *exponential order* b if $e^{-bx}f(x)$ is bounded on $[0, \infty)$; i.e., if there is a number B such that $e^{-bx}|f(x)| \le B$ for all $x \ge 0$.

THEOREM 1. *If f is of exponential order b, then $\hat{f}(s)$ exists for all $s > b$.*

Proof. If $s > b$, then $s = b + p$ for some $p > 0$, and

$$e^{-sx}|f(x)| = e^{-bx}|f(x)|e^{-px} \le Be^{-px}.$$

Since $p > 0$, the integral

$$\int_0^\infty Be^{-px}\, dx$$

converges, and therefore the integral for $\hat{f}(s)$ converges by Theorem 1, Section 5–1.

We restate Theorem 3 of the last section and its corollary here in terms of our standard hypotheses of continuity and exponential order.

THEOREM 2. *If f and f' are continuous and f is of exponential order b, then $\mathcal{L}(f'(x))$ exists for $s > b$, and*

$$\mathcal{L}(f'(x)) = s\mathcal{L}(f(x)) - f(0).$$

Proof. If f is of exponential order b, then $\lim_{x \to \infty} e^{-sx}f(x) = 0$ for $s > b$ (Problem 1), and this is the condition of Theorem 3, Section 5–2.

COROLLARY. *If f is of exponential order b and*

$$F(x) = \int_0^x f(t)\, dt,$$

then F is of exponential order b and

$$\mathcal{L}(F(x)) = (1/s)\mathcal{L}(f(x)).$$

Proof. By the corollary of Theorem 3, Section 5–2, it is sufficient to show that F is of exponential order b. Write

$$F(x) = \int_0^x f(t)\, dt = \int_0^x e^{bt}e^{-bt}f(t)\, dt,$$

and conclude that

$$|F(x)| \le \int_0^x Be^{bt}\, dt = \frac{B}{b}\,(e^{bx} - 1).$$

Therefore, $e^{-bx}|F(x)| \le B/|b|$, and F is also of exponential order b.

If the functions f, f', f'' etc., are all continuous and of exponential order b, then by repeated applications of Theorem 2, we obtain the following basic formulas for $s > b$:

$$\mathcal{L}(f'(x)) = s\hat{f}(s) - f(0),$$
$$\mathcal{L}(f''(x)) = s^2\hat{f}(s) - sf(0) - f'(0), \tag{1}$$
$$\mathcal{L}(f'''(x)) = s^3\hat{f}(s) - s^2 f(0) - sf'(0) - f''(0), \quad \text{etc.}$$

THEOREM 3. *If f is of exponential order b, then $e^{-ax}f(x)$ is of exponential order $b - a$, and for $s > b - a$, we have*

$$\mathcal{L}(e^{-ax}f(x)) = \hat{f}(s + a). \tag{2}$$

Proof. Problem 2.

As an example of (2), consider the formula (see Problem 4, Section 5–2)

$$\mathcal{L}(\sin bx) = \frac{b}{s^2 + b^2}, \quad (s > 0).$$

From this and (2) we have immediately

$$\mathcal{L}(e^{-ax}\sin bx) = \frac{b}{(s + a)^2 + b^2}, \quad (s > -a).$$

Similarly, from

$$\mathcal{L}(x) = \frac{1}{s^2}, \quad (s > 0),$$

we get (with $a = -1$)

$$\mathcal{L}(xe^x) = \frac{1}{(s - 1)^2}, \quad (s > 1).$$

Formulas (1) describe what happens to the transform when the object function is differentiated. The next theorem describes what happens to the object function when the transform is differentiated. More precisely, we show that each transform $\hat{f}$ is differentiable, that its derivative $\hat{f}'$ is again a transform, and that

$$\hat{f}'(s) = \mathcal{L}(-xf(x)). \tag{3}$$

The proof of these facts depends on the following lemma.

LEMMA 1. *If f is of exponential order b, and n is a positive integer, then* $\mathcal{L}(x^n f(x))$ *exists for* $s > b$, *and the integral for this transform converges uniformly for* $s \geq b_0$, *for any* $b_0 > b$.

Proof. Problem 3.

Notice that the lemma above implies that the integral for $\mathcal{L}(x^n f(x))$ converges uniformly on some interval around s, for each $s > b$.

THEOREM 4. *If f is of exponential order b, then* $\hat{f}$ *has derivatives of all orders for* $s > b$, *and*

$$\hat{f}'(s) = \mathcal{L}(-xf(x)),$$
$$\hat{f}''(s) = \mathcal{L}(x^2 f(x)), \tag{4}$$
$$\hat{f}'''(s) = \mathcal{L}(-x^3 f(x)), \quad \text{etc.}$$

Proof. The formulas (4) are obtained by repeatedly differentiating the integral formula for $\hat{f}(s)$ under the integral sign. For example,

$$\hat{f}(s) = \int_0^\infty e^{-sx} f(x)\, dx, \tag{5}$$

$$\hat{f}'(s) = \int_0^\infty e^{-sx}(-x)f(x)\, dx. \tag{6}$$

Differentiation under the integral sign in (5) is justified by Theorem 4, Section 5–1, since by the lemma we know that the resulting integral (6) converges uniformly on some interval around each $s > b$.

As an example of how new transforms can be obtained from (4), consider the formula (see Problem 3, Section 5–1)

$$\mathcal{L}(e^{ax}) = \frac{1}{s - a}, \quad (s > a).$$

By differentiating the right side, we get

$$\mathcal{L}(xe^{ax}) = \frac{1}{(s - a)^2}, \quad (s > a),$$

and in general

$$\mathcal{L}(x^{n-1} e^{ax}) = \frac{(n - 1)!}{(s - a)^n}, \quad (s > a).$$

In a similar way, we can start with

$$\mathcal{L}(1) = \frac{1}{s}, \quad (s > 0)$$

and obtain the formulas

$$\mathcal{L}(x^n) = \frac{n!}{s^{n+1}}, \quad (s > 0).$$

Since our purpose is to solve linear differential equations with constant coefficients by means of transforms, we need to know that the solutions we seek always have transforms. This question is settled by the following lemmas and theorem.

LEMMA 2. *If $f_1, \ldots, f_n$ are of exponential orders $b_1, \ldots, b_n$, respectively, then $f_1 + \cdots + f_n$ is of exponential order $b = \max\{b_1, \ldots, b_n\}$.*

Proof. Problem 4.

LEMMA 3. *If $q, q', q'', \ldots, q^{(k)}$ are continuous and of exponential order b, and y is a solution of*

$$(D - r)y = q, \tag{7}$$

then $y, y', \ldots, y^{(k+1)}$ are continuous and of exponential order $b' = \max\{b, r\}$.

Proof. Any solution y of (7) can be written

$$y(x) = ce^{rx} + e^{rx} \int_0^x e^{-rt} q(t) \, dt. \tag{8}$$

The solution y will have the required properties if the integral

$$u(x) = \int_0^x e^{-rt} q(t) \, dt \tag{9}$$

has $k + 1$ continuous derivatives which are of exponential order $b - r$. The function u itself is of order $b - r$, by the corollary of Theorem 2, since the integrand in (9) is of order $b - r$. Also, u is differentiable (hence continuous), and its derivatives are given by

$$u'(x) = e^{-rx} q(x),$$
$$u''(x) = e^{-rx}[q'(x) - rq(x)], \tag{10}$$
$$u'''(x) = e^{-rx}[q''(x) - 2rq'(x) + r^2 q(x)], \quad \text{etc.}$$

From (10) it is clear that u, and hence y, has one more continuous derivative than q. From Lemma 2 and (10) we see that $u', u'', \ldots, u^{(k+1)}$ are of exponential order $b - r$, since the terms in brackets are of order b.

If r is a complex number, we modify the proof above by considering the real part of r, and the lemma holds for real or complex numbers r.

THEOREM 5. *If q is continuous and of exponential order, and y is a solution of*

$$y^{(n)} + p_{n-1}y^{(n-1)} + \cdots + p_1 y' + p_0 y = q, \tag{11}$$

where $p_0, p_1, \ldots, p_{n-1}$ are constants, then $y, y', \ldots, y^{(n)}$ are continuous, and of exponential order.

Proof. We write Eq. (11) in the form

$$(D - r_1)(D - r_2) \cdots (D - r_n)y = q \tag{12}$$

and assume that y is a given solution of (12). Define functions $y_1, y_2, \ldots, y_{n-1}$ by

$$(D - r_2)(D - r_3) \cdots (D - r_n)y = y_1,$$
$$(D - r_3) \cdots (D - r_n)y = y_2,$$
$$\vdots$$
$$(D - r_n)y = y_{n-1}.$$

From these equations, we have

$$(D - r_1)y_1 = q,$$
$$(D - r_2)y_2 = y_1,$$
$$\vdots \tag{13}$$
$$(D - r_{n-1})y_{n-1} = y_{n-2},$$
$$(D - r_n)y = y_{n-1}.$$

By Lemma 3 and the first of equations (13), y_1 and y_1' are continuous and of exponential order. From the second equation it then follows that y_2, y_2', and y_2'' are continuous and of exponential order. Continuing in this way, we see that $y_{n-1}, y_{n-1}', \ldots, y_{n-1}^{(n-1)}$ are continuous and of exponential order. The last equation in (13) and Lemma 3 then guarantee that $y, y', \ldots, y^{(n)}$ are continuous and of exponential order.

COROLLARY. *If q is continuous and of exponential order, and y is a solution of (11), then $\mathcal{L}(y(x)), \mathcal{L}(y'(x)), \ldots, \mathcal{L}(y^{(n)}(x))$ exist for all sufficiently large values of s.*

The fact that $\mathcal{L}$ is a *one-to-one* operator is fundamental in our applications. Using the following theorem, we will prove this next.

THEOREM 6 (Weierstrass polynomial approximation theorem). *If u is continuous on $[0, 1]$ and $\epsilon > 0$, then there is a polynomial P such that $|u(x) - P(x)| < \epsilon$ for all x in $[0, 1]$.*

Theorem 6 is one of the most useful theorems of analysis. Although the statement of the theorem is quite simple, the proof is complicated and will be omitted.

COROLLARY. *If u is continuous on $[0, 1]$ and $\int_0^1 x^n u(x)\, dx = 0$ for $n = 0, 1, 2, \ldots$, then $u(x) \equiv 0$.*

Proof. Let ϵ be any positive number, and let P be a polynomial such that $|u(x) - P(x)| < \epsilon$ for all x in $[0, 1]$. By the hypothesis,

$$\int_0^1 P(x)u(x)\, dx = 0,$$

and hence

$$\int_0^1 u(x)^2\, dx = \int_0^1 u(x)[u(x) - P(x)]\, dx \leq \epsilon \int_0^1 |u(x)|\, dx.$$

Since ϵ is arbitrary, the integral of $u(x)^2$ is zero, and hence $u(x) \equiv 0$.

THEOREM 7 ($\mathcal{L}$ is one-to-one). *If f and g are continuous on $[0, \infty)$ and $\hat{f}(s) = \hat{g}(s)$ for all $s \geq s_0$, then $f(x) \equiv g(x)$.*

Proof. It is sufficient (Problem 6) to show that if h is any continuous function such that $\hat{h}(s) = 0$ for $s \geq s_0$, then $h(x) \equiv 0$. Assume therefore that $\hat{h}(s) = 0$ for $s \geq s_0$, and in particular that $\hat{h}(s_0 + n) = 0$ for $n = 0, 1, 2, \ldots$:

$$\hat{h}(s_0 + n) = \int_0^\infty e^{-nx} e^{-s_0 x} h(x)\, dx = 0. \tag{14}$$

Define the function v as follows:

$$v(x) = \int_0^x e^{-s_0 t} h(t)\, dt. \tag{15}$$

Note that v is continuous on $[0, \infty)$, with $v(0) = 0$ and $\lim_{x \to \infty} v(x) = \hat{h}(s_0) = 0$. Now integrate (14) by parts, with $u = e^{-nx}$ and $dv = e^{-s_0 x} h(x)\, dx$, so that v is given by (15). We get

$$0 = \hat{h}(s_0 + n) = \lim_{R \to \infty} \left[e^{-nR} v(R) + n \int_0^R e^{-nx} v(x)\, dx \right]$$
$$= n \int_0^\infty e^{-nx} v(x)\, dx;$$

that is, for $n = 0, 1, 2, \ldots$, we have

$$\int_0^\infty e^{-nx} v(x)\, dx = 0. \tag{16}$$

We make a change of variable in (16), with $e^{-x} = t, e^{-nx} = t^n, x = \ln t^{-1}$, $dx = -(1/t)\,dt$, and let $u(t) = v(\ln t^{-1}) = v(x)$. As x ranges between 0 and ∞, t ranges between 1 and 0. If we let $u(0) = 0$, then u is continuous on $[0, 1]$, since $v(x) \to 0$ as $x \to \infty$ $(t \to 0)$. With this change of variable, (16) becomes

$$\int_0^1 t^{n-1}u(t)\,dt = 0.$$

Therefore $u(t) \equiv 0$ on $[0, 1]$, and $v(x) \equiv 0$ on $[0, \infty)$. Hence $v'(x) = e^{-s_0 x}h(x) \equiv 0$, and $h(x) \equiv 0$.

Since $\mathcal{L}$ is one-to-one, we can define the inverse operator $\mathcal{L}^{-1}$ which maps transforms onto object functions. In other words,

$$\mathcal{L}^{-1}(\phi(s)) = f(x)$$

if and only if $\mathcal{L}(f(x)) = \phi(s)$. It is easy to show (Problem 12) that $\mathcal{L}^{-1}$ is also a linear operator; i.e., that

$$\mathcal{L}^{-1}(a\phi(s) + b\psi(s)) = a\mathcal{L}^{-1}(\phi(s)) + b\mathcal{L}^{-1}(\psi(s)). \tag{17}$$

In some applications it is convenient to allow functions q in (11) which have discontinuities. For example, such a function would arise in the discussion of an electric circuit in which a constant voltage was applied at some time $t_0 > 0$. Although the theorems of this section are stated and proved for continuous functions, they extend with only minor changes to functions with jump discontinuities at isolated points.

PROBLEMS

1. Show that if f is of exponential order b, then $\lim_{x\to\infty} e^{-sx}f(x) = 0$ for all $s > b$.

2. Prove Theorem 3.

*3. Prove Lemma 1. [*Hint:* Assume that $b_0 > b$, and let $p = b_0 - b > 0$. Show that if $s \geq b_0$, then $e^{-sx}x^n|f(x)| \leq Be^{-px}x^n$. Then use Theorem 3, Section 5-1, and Problem 2, Section 5-2.]

4. Prove Lemma 2.

5. Explain the difference between the Weierstrass polynomial approximation theorem and the *false* statement that every continuous function has a power series expansion.

6. (See Theorem 7.) Show that $\hat{f}(s) \equiv \hat{g}(s)$ implies $f(x) \equiv g(x)$ if and only if $\hat{h}(s) \equiv 0$ implies $h(x) \equiv 0$.

7. Show that $\mathcal{L}(e^{-x^2})$ exists for every value of s. [*Hint:* Show that $e^{-sx}e^{-x^2} \leq e^{-x}$ if $x \geq |s| + 1$, and use Theorem 1, Section 5-1.]

8. Show that $\mathcal{L}(e^{x^2})$ does not exist for any value of s. [*Hint:* Show that for every s, $\lim_{x\to\infty} e^{-sx}e^{x^2} = \infty$.]

9. For any number $a > 0$, let f_a be defined as follows:

$$f_a(x) = \begin{cases} f(x - a) & \text{if } x \geq a, \\ 0 & \text{if } 0 \leq x < a. \end{cases}$$

Show that $\mathcal{L}\big(f_a(x)\big) = \hat{f}(s + a)$.

10. Let f be the function which is zero on $[0, 1)$ and one on $[1, \infty)$. Find $\mathcal{L}\big(f(x)\big)$.

11. All the formulas of Table 2, Section 5-4, can be derived from Formulas 1 and 9 and the theorems of this section. Without using the definition, derive

(a) Formulas 2 through 4.   (b) Formulas 5 through 8.
(c) Formulas 10 through 12.   (d) Formulas 13 through 15.

12. Show that $\mathcal{L}^{-1}$ is a linear operator; i.e., verify formula (17).

5-4 Solution of equations by transforms. Table 1 is a list of the operational formulas which were proved in Section 5-3 under appropriate assumptions, and one formula, (E), which is discussed below. Table 2 is a short table of Laplace transforms of specific elementary functions. The object functions listed in Table 2 are all of exponential order, and hence each of the transforms is defined for all sufficiently large values of s. We start with some examples to show how such a table is used to obtain solutions of linear equations with constant coefficients and specified initial conditions.

EXAMPLE 1. $y'' + 4y = x + \sin 2x$, $y(0) = y'(0) = 0$.

Let y be the solution of this equation, and $\hat{y}$ be its transform. For simplicity, we will write $\hat{y}$ instead of $\hat{y}(s)$ in the transformed equation in the same way that y, y'' are used for $y(x)$, $y''(x)$ in the differential equation

TABLE 1

A	$\mathcal{L}\big(f'(x)\big) = s\hat{f}(s) - f(0)$
B	$\mathcal{L}\left(\int_0^x f(t)\, dt\right) = \dfrac{1}{s}\,\hat{f}(s)$
C	$\mathcal{L}\big(e^{-ax}f(x)\big) = \hat{f}(s + a)$
D	$\mathcal{L}\big(-xf(x)\big) = \hat{f}'(s)$
E	$\mathcal{L}\big(f(x)*g(x)\big) = \hat{f}(s)\hat{g}(s)$

TABLE 2

	$\hat{f}(s)$	$f(x)$
1	$\dfrac{1}{s}$	1
2	$\dfrac{1}{s^n}$	$\dfrac{1}{(n-1)!}\,x^{n-1} \quad (n=1,2,3,\ldots)$
3	$\dfrac{1}{s-a}$	e^{ax}
4	$\dfrac{1}{(s-a)^n}$	$\dfrac{1}{(n-1)!}\,x^{n-1}e^{ax} \quad (n=1,2,3,\ldots)$
5	$\dfrac{1}{s^2-a^2}$	$\dfrac{1}{a}\sinh ax$
6	$\dfrac{s}{s^2-a^2}$	$\cosh ax$
7	$\dfrac{1}{(s-a)(s-b)} \quad (a \neq b)$	$\dfrac{1}{a-b}\,[e^{ax}-e^{bx}]$
8	$\dfrac{s}{(s-a)(s-b)} \quad (a \neq b)$	$\dfrac{1}{a-b}\,[ae^{ax}-be^{bx}]$
9	$\dfrac{1}{s^2+a^2}$	$\dfrac{1}{a}\sin ax$
10	$\dfrac{s}{s^2+a^2}$	$\cos ax$
11	$\dfrac{1}{(s-a)^2+b^2}$	$\dfrac{1}{b}\,e^{ax}\sin bx$
12	$\dfrac{s-a}{(s-a)^2+b^2}$	$e^{ax}\cos bx$
13	$\dfrac{s}{(s^2+a^2)^2}$	$\dfrac{1}{2a}\,x\sin ax$
14	$\dfrac{s^2-a^2}{(s^2+a^2)^2}$	$x\cos ax$
15	$\dfrac{1}{(s^2+a^2)^2}$	$\dfrac{1}{2a^3}\,[\sin ax - ax\cos ax]$

itself. The transform $\hat{y}$ must satisfy

$$s^2\hat{y} - sy(0) - y'(0) + 4\hat{y} = \frac{1}{s^2} + \frac{2}{s^2 + 4},$$

or, since $y(0) = y'(0) = 0$,

$$\hat{y}(s^2 + 4) = \frac{1}{s^2} + \frac{2}{s^2 + 4},$$

$$\hat{y} = \frac{1}{s^2(s^2 + 4)} + \frac{2}{(s^2 + 4)^2}.$$

Hence

$$y = \mathcal{L}^{-1}\left\{\frac{1}{s^2(s^2 + 4)}\right\} + 2\mathcal{L}^{-1}\left\{\frac{1}{(s^2 + 4)^2}\right\}.$$

From Table 2 (#15), we find

$$2\mathcal{L}^{-1}\left\{\frac{1}{(s^2 + 4)^2}\right\} = \tfrac{1}{8}[\sin 2x - 2x\cos 2x]. \tag{1}$$

Since $1/s^2(s^2 + 4)$ does not appear in the table, we break this expression into partial fractions:

$$\frac{1}{s^2(s^2 + 4)} = \frac{A}{s} + \frac{B}{s^2} + \frac{Cs + D}{s^2 + 4}. \tag{2}$$

Equation (2) is an identity if

$$1 \equiv (A + C)s^3 + (B + D)s^2 + 4As + 4B.$$

Hence $B = \tfrac{1}{4}, A = 0, D = -\tfrac{1}{4}, C = 0$, and

$$\frac{1}{s^2(s^2 + 4)} = \frac{1}{4}\frac{1}{s^2} - \frac{1}{4}\frac{1}{s^2 + 4}.$$

From the table (#2, #9), we get

$$\mathcal{L}^{-1}\left\{\frac{1}{s^2(s^2 + 4)}\right\} = \tfrac{1}{4}x - \tfrac{1}{4}\tfrac{1}{2}\sin 2x. \tag{3}$$

Finally, from (1) and (3), we have

$$y = \tfrac{1}{8}[\sin 2x - 2x\cos 2x] + \tfrac{1}{4}x - \tfrac{1}{8}\sin 2x$$
$$= \tfrac{1}{4}x - \tfrac{1}{4}x\cos 2x.$$

EXAMPLE 2. $y'' + 2y' + 2y = 2e^{-x} \cos x, \quad y(0) = 2, \quad y'(0) = -2.$

Taking the transform of both sides, using #12, we get

$$s^2\hat{y} - 2s + 2 + 2(s\hat{y} - 2) + 2\hat{y} = \frac{2(s + 1)}{(s + 1)^2 + 1}.$$

Collecting terms and simplifying, we obtain

$$\hat{y}(s^2 + 2s + 2) = \frac{2(s + 1)}{(s + 1)^2 + 1} + 2(s + 1),$$

$$\hat{y} = \frac{2(s + 1)}{[(s + 1)^2 + 1]^2} + \frac{2(s + 1)}{(s + 1)^2 + 1}. \tag{4}$$

Note that the first term on the right of (4) can be written

$$\frac{2(s + 1)}{[(s + 1)^2 + 1]^2} = \frac{d}{ds} \frac{-1}{(s + 1)^2 + 1}.$$

Therefore, using (D), then #11 and #12, we have

$$y = -x\mathcal{L}^{-1}\left\{\frac{-1}{(s + 1)^2 + 1}\right\} + 2\mathcal{L}^{-1}\left\{\frac{s + 1}{(s + 1)^2 + 1}\right\}$$

$$= xe^{-x} \sin x + 2e^{-x} \cos x.$$

Instead of using (D) as above, we can use (C), and then #13, to obtain

$$\mathcal{L}^{-1}\left\{\frac{2(s + 1)}{[(s + 1)^2 + 1]^2}\right\} = 2e^{-x}\mathcal{L}^{-1}\left\{\frac{s}{[s^2 + 1]^2}\right\}$$

$$= 2e^{-x}\tfrac{1}{2}x \sin x.$$

Because of the form of the entries in Table 2, it is usually necessary to break a transform into partial fractions before we can identify the inverse transform. Next we give a simple method for effecting the partial fraction decomposition for a rational function $P(s)/Q(s)$ when $Q(s)$ is a product of distinct linear factors. Suppose that $r_1, \ldots, r_n$ are distinct numbers, that

$$Q(s) = (s - r_1) \cdots (s - r_n),$$

and that P is a polynomial of degree less than n. Then there are numbers $A_1, \ldots, A_n$ such that

$$\frac{P(s)}{Q(s)} = \frac{A_1}{s - r_1} + \cdots + \frac{A_n}{s - r_n}. \tag{5}$$

If we multiply both sides of (5) by $s - r_i$, and let s approach r_i, only A_i remains on the right, and hence

$$A_i = \lim_{s \to r_i} \frac{(s - r_i)P(s)}{Q(s)}. \tag{6}$$

Since $Q(r_i) = 0$, we can write (6) in the form

$$A_i = \lim_{s \to r_i} \frac{P(s)}{\dfrac{Q(s) - Q(r_i)}{s - r_i}} = \frac{P(r_i)}{Q'(r_i)}.$$

Therefore the decomposition (5) can be written

$$\frac{P(s)}{Q(s)} = \frac{P(r_1)/Q'(r_1)}{s - r_1} + \cdots + \frac{P(r_n)/Q'(r_n)}{s - r_n}.$$

EXAMPLE 3. Express in partial fractions

$$\frac{s^2 - 2s + 2}{s^3 - s^2 - 4s + 4}.$$

Here $Q(s) = s^3 - s^2 - 4s + 4 = (s - 1)(s - 2)(s + 2)$, and $Q'(s) = 3s^2 - 2s - 4$. Hence $P(1)/Q'(1) = 1/(-3) = -\frac{1}{3}$, $P(2)/Q'(2) = \frac{2}{4} = \frac{1}{2}$, $P(-2)/Q'(-2) = \frac{10}{12} = \frac{5}{6}$, and

$$\frac{s^2 - 2s + 2}{s^3 - s^2 - 4s + 4} = -\frac{1}{3}\frac{1}{s - 1} + \frac{1}{2}\frac{1}{s - 2} + \frac{5}{6}\frac{1}{s + 2}.$$

EXAMPLE 4. $y'' - 2y' - 3y = e^x,$ $y(0) = 1,$ $y'(0) = 1.$

Taking the transform of both sides and simplifying, we get

$$s^2\hat{y} - s - 1 - 2(s\hat{y} - 1) - 3\hat{y} = \frac{1}{s - 1},$$

$$\hat{y}(s^2 - 2s - 3) = \frac{1}{s - 1} + s - 1 = \frac{s^2 - 2s + 2}{s - 1},$$

$$\hat{y} = \frac{s^2 - 2s + 2}{(s - 1)(s - 3)(s + 1)}.$$

Here $P(s) = s^2 - 2s + 2$, $Q(s) = s^3 - 3s^2 - s + 3$, and $Q'(s) = 3s^2 - 6s - 1$. Hence $P(1)/Q'(1) = -\frac{1}{4}$, $P(3)/Q'(3) = \frac{5}{8}$, $P(-1)/Q'(-1) = \frac{5}{8}$, and

$$\hat{y} = \frac{s^2 - 2s + 2}{(s - 1)(s - 3)(s + 1)} = -\frac{1}{4}\frac{1}{s - 1} + \frac{5}{8}\frac{1}{s - 3} + \frac{5}{8}\frac{1}{s + 1}.$$

Therefore

$$y = -\tfrac{1}{4}e^x + \tfrac{5}{8}e^{3x} + \tfrac{5}{8}e^{-x}.$$

We continually use the linear properties of $\mathcal{L}$—that is, we continually use the fact that $\mathcal{L}$ is a mapping which preserves the operations of addition of functions and multiplication of a function by a constant. There is a type of multiplication for object functions, denoted $f*g$, for which $\mathcal{L}$ is also a multiplicative mapping, in the sense that

$$\mathcal{L}[(f*g)(x)] = \mathcal{L}(f(x))\mathcal{L}(g(x)).$$

The product on the right above is the usual pointwise product of two functions. The "product" $f*g$, called the *convolution* of f and g, is defined by

$$(f*g)(x) = \int_0^x f(x - t)g(t)\, dt. \tag{7}$$

For convenience in discussing the convolution of particular formulas, we will usually use the slightly improper notation $f(x)*g(x)$ instead of $(f*g)(x)$.

We give here without proof a formal statement of the multiplicative property of $\mathcal{L}$ with respect to convolution.

THEOREM 1. *If f and g are continuous on $[0, \infty)$ and of exponential order b, then $f*g$ is continuous and of exponential order b, and for all $s > b$*

$$\mathcal{L}[f(x)*g(x)] = \mathcal{L}(f(x))\mathcal{L}(g(x)). \tag{8}$$

Convolution has the following properties, which justify considering this operation as a type of product for functions.

$$\begin{aligned}
f*g &= g*f, \\
(f*g)*h &= f*(g*h), \\
f*(g + h) &= f*g + f*h, \\
(cf)*g &= f*(cg) = c(f*g).
\end{aligned} \tag{9}$$

The formulas above can be derived directly from the definition (7), but follow most easily from (8) and the fact that $\mathcal{L}$ is one-to-one. For example, from (8) we have

$$\mathcal{L}[f(x)*g(x)] = \mathcal{L}(f(x))\mathcal{L}(g(x))$$

and

$$\mathcal{L}[g(x)*f(x)] = \mathcal{L}(g(x))\mathcal{L}(f(x)).$$

The right sides of the two equations above are obviously equal, so

$$\mathcal{L}[f(x)*g(x)] = \mathcal{L}[g(x)*f(x)].$$

Since $\mathcal{L}$ is a one-to-one operator, it follows that

$$f(x)*g(x) = g(x)*f(x).$$

We list next some examples of the convolution of two functions and the parallel statements for their transforms.

$$1*f(x) = \int_0^x f(t)\, dt,$$
$$\mathcal{L}(1*f(x)) = \mathcal{L}(1)\mathcal{L}(f(x)) = \frac{1}{s}\,\mathcal{L}(f(x)). \tag{10}$$

$$x*x = \int_0^x (x - t)t\, dt = \tfrac{1}{6}x^3,$$
$$\mathcal{L}(x*x) = \mathcal{L}(x)\mathcal{L}(x) = \frac{1}{s^2}\cdot\frac{1}{s^2} = \frac{1}{s^4}. \tag{11}$$

$$x* \sin x = \int_0^x (x - t)\sin t\, dt = x - \sin x,$$
$$\mathcal{L}(x* \sin x) = \mathcal{L}(x)\mathcal{L}\,(\sin x) = \frac{1}{s^2}\frac{1}{s^2 + 1} = \frac{1}{s^2} - \frac{1}{s^2 + 1}. \tag{12}$$

$$x*e^x = \int_0^x (x - t)e^t\, dt = e^x - x - 1,$$
$$\mathcal{L}(x*e^x) = \mathcal{L}(x)\mathcal{L}(e^x) = \frac{1}{s^2}\cdot\frac{1}{s - 1} = \frac{1}{s - 1} - \frac{1}{s^2} - \frac{1}{s}. \tag{13}$$

$$\sin x* \cos x = \int_0^x \sin\,(x - t)\cos t\, dt = \tfrac{1}{2}x \sin x,$$
$$\mathcal{L}\,(\sin x* \cos x) = \mathcal{L}\,(\sin x)\mathcal{L}\,(\cos x) = \frac{1}{s^2 + 1}\cdot\frac{s}{s^2 + 1} = \frac{s}{(s^2 + 1)^2}. \tag{14}$$

EXAMPLE 5. $y'' + y = \cos x,$ $y(0) = y'(0) = 0.$

The transformed equation is

$$s^2\hat{y} + \hat{y} = \frac{s}{s^2 + 1},$$

or

$$\hat{y} = \frac{1}{s^2 + 1}\cdot\frac{s}{s^2 + 1} = \mathcal{L}\,(\sin x)\mathcal{L}\,(\cos x).$$

It follows from (8) and (14) that

$$y = \sin x* \cos x = \tfrac{1}{2}x \sin x.$$

EXAMPLE 6. $y'' - 5y' + 6y = e^{-2x}$, $\quad y(0) = 0$, $\quad y'(0) = -2$.

The transformed equation is

$$\hat{y}(s^2 - 5s + 6) = \frac{1}{s + 2} - 2,$$

or

$$\hat{y} = \frac{1}{(s + 2)(s - 2)(s - 3)} - \frac{2}{(s - 2)(s - 3)}. \tag{15}$$

From Table 2 (#7), we get

$$\mathcal{L}^{-1}\left\{\frac{1}{(s - 2)(s - 3)}\right\} = e^{3x} - e^{2x}. \tag{16}$$

Therefore

$$\mathcal{L}^{-1}\left\{\frac{1}{s + 2} \cdot \frac{1}{(s - 2)(s - 3)}\right\} = e^{-2x} * [e^{3x} - e^{2x}]$$

$$= \int_0^x e^{-2(x-t)}[e^{3t} - e^{2t}]\,dt$$

$$= e^{-2x}\int_0^x e^{2t}[e^{3t} - e^{2t}]\,dt$$

$$= e^{-2x}[\tfrac{1}{5}e^{5x} - \tfrac{1}{5} - \tfrac{1}{4}e^{4x} + \tfrac{1}{4}]$$

$$= \tfrac{1}{5}e^{3x} - \tfrac{1}{4}e^{2x} - \tfrac{1}{20}e^{-2x}. \tag{17}$$

Combining the results of (15), (16), and (17), we have

$$y = \mathcal{L}^{-1}\left\{\frac{1}{(s + 2)(s - 2)(s - 3)}\right\} - 2\mathcal{L}^{-1}\left\{\frac{1}{(s - 2)(s - 3)}\right\}$$

$$= \tfrac{1}{5}e^{3x} - \tfrac{1}{4}e^{2x} - \tfrac{1}{20}e^{-2x} - 2[e^{3x} - e^{2x}]$$

$$= -\tfrac{9}{5}e^{3x} + \tfrac{7}{4}e^{2x} - \tfrac{1}{20}e^{-2x}.$$

PROBLEMS

Solve the following equations using transforms.

1. $y' - y = e^x$, $y(0) = 1$
2. $y' = x$, $y(0) = 1$
3. $y'' - 3y' + 2y = e^{-x}$, $y(0) = -1$, $y'(0) = 1$
4. $y'' + 9y = \sin 3x$, $y(0) = 1$, $y'(0) = 0$
5. $y'' + 2y' + y = 1 + e^x$, $y(0) = 1$, $y'(0) = 0$
6. $y'' + y = x + e^x$, $y(0) = 2$, $y'(0) = 1$
7. $y'' - 3y' + 2y = 2e^{3x}$, $y(0) = 2$, $y'(0) = 3$

8. $y''' - y' = e^{2x}$, $y(0) = y'(0) = y''(0) = 0$
9. $y'' + y' = 3x^2 - 6$, $y(0) = 0$, $y'(0) = 1$
10. $y''' - y' = 2 \sin x$, $y(0) = y'(0) = y''(0) = 0$
11. $y'' + 2y' + 2y = 0$, $y(0) = y'(0) = 1$
12. $y'' - 2y' + y = e^x \sin x$, $y(0) = y'(0) = 0$
13. $y'' - 4y' + 5y = e^{2x} \cos x$, $y(0) = y'(0) = 0$

14. Verify the convolution formulas in (10), (11), (12), (13), and (14). Check in each case that $\mathcal{L}[f(x)*g(x)] = \mathcal{L}(f(x))\mathcal{L}(g(x))$.

15. Make the change of variable $t = x - u$, $dt = -du$ in (7) and verify that $f*g = g*f$.

16. (a) Show that $f(ax)*g(ax) = (1/a)(f*g)(ax)$.
    (b) Use (a) and (14) to conclude that $(\sin 2x)*(\cos 2x) = \frac{1}{2}x \sin 2x$.
    (c) Use (b) to solve $y'' + 4y = \cos 2x$, $y(0) = y'(0) = 0$.

17. (a) Use the definition (7) to show that

$$(\sin ax)* (\sin ax) = \frac{1}{2a} \sin ax - \tfrac{1}{2}x \cos ax.$$

    (b) Derive the formula of (a) using transforms (#9 and #15 of Table 2).

18. (a) Use the definition (7) to show that

$$(\cos ax)* (\cos ax) = \tfrac{1}{2}x \cos ax + \frac{1}{2a} \sin ax.$$

    (b) Derive the formula of (a) using transforms (#10, #14, and #15 of Table 2).

Answers

1. $y = (1 + x)e^x$
2. $y = 1 + \frac{1}{2}x^2$
3. $y = \frac{7}{3}e^{2x} - \frac{7}{2}e^x + \frac{1}{6}e^{-x}$
4. $y = \frac{1}{18} \sin 3x + \cos 3x - \frac{1}{6}x \cos 3x$
5. $y = 1 + \frac{1}{4}e^x - \frac{1}{4}e^{-x} - \frac{1}{2}xe^{-x}$
6. $y = x + \frac{1}{2}e^x + \frac{3}{2} \cos x - \frac{1}{2} \sin x$
7. $y = 2e^x - e^{2x} + e^{3x}$
8. $y = \frac{1}{6}e^{2x} - \frac{1}{2}e^x - \frac{1}{6}e^{-x} + \frac{1}{2}$
9. $y = 1 - 3x^2 + x^3 - e^{-x}$
10. $y = -2 + \cos x + \frac{1}{2}e^x + \frac{1}{2}e^{-x}$
11. $y = e^{-x} \cos x + 2e^{-x} \sin x$
12. $y = xe^x - e^x \sin x$
13. $y = \frac{1}{2}xe^{2x} \sin x$
16. (c) $y = \frac{1}{4}x \sin 2x$

CHAPTER 6

PICARD'S EXISTENCE THEOREM

6–1 Review. We will start this chapter with a review of some of the basic definitions and theorems from calculus which will be used in the existence proof.

A function f of one variable is *continuous at a* if for every $\epsilon > 0$ there is a $\delta > 0$ such that $|f(x) - f(a)| < \epsilon$ whenever $|x - a| < \delta$. The function f is *continuous on an interval I* if f is continuous at each point of I. A function F of two variables is *continuous at (a, b)* if for every $\epsilon > 0$ there is a $\delta > 0$ such that $|F(x, y) - F(a, b)| < \epsilon$ whenever $|x - a| < \delta$ and $|y - b| < \delta$. The function F is *continuous on a set of points* (e.g., a square) S if F is continuous at each point of S.

THEOREM 1. *A function of one variable which is continuous on a closed interval is bounded on that interval, and similarly a function of two variables which is continuous on a closed square (including the boundary lines) is bounded.*

We will use the notation

$$\max_{t \text{ in } [a, b]} |f(t)|$$

for the *least* upper bound of the numbers $|f(t)|$ for t in $[a, b]$.

The student is familiar with the basic properties of the definite integral

$$\int_a^b f(t)\, dt. \tag{1}$$

We recall that there are many functions f for which (1) is *not* defined, but that (1) is defined for any function f which is continuous on $[a, b]$ (or $[b, a]$, if $b < a$). The following is essentially the fundamental theorem of calculus.

THEOREM 2. *If f is continuous on $[a, b]$ and g is defined on $[a, b]$ by*

$$g(x) = \int_a^x f(t)\, dt,$$

then g is differentiable, and $g'(x) = f(x)$ for x in $[a, b]$.

We will also use the following two integral inequalities.

THEOREM 3. *If f is continuous on [a, b] (or [b, a]), then*

$$\left| \int_a^b f(t)\, dt \right| \le \left| \int_a^b |f(t)|\, dt \right| \tag{2}$$

and

$$\left| \int_a^b f(t)\, dt \right| \le |b - a| \max_{t \text{ in } [a,b]} |f(t)|. \tag{3}$$

The Picard method "constructs" a solution to the differential equation in question as the limit of a sequence of functions. We turn to some facts about sequences.

The sequence (of numbers) $\{a_n\}$ *converges to a* (denoted $\lim_{n\to\infty} a_n = a$) if for every positive number ϵ there is an integer N such that $|a_n - a| < \epsilon$ if $n \ge N$.

If $\{y_n\}$ is a sequence of functions defined on a given interval I, then we can regard this as a family of sequences of numbers; i.e., for each fixed x in I, we have the sequence $\{y_n(x)\}$ of numbers. The limit of $\{y_n(x)\}$, if it exists, will of course depend on x, and hence a function y is defined by $y(x) = \lim_{n\to\infty} y_n(x)$. If the sequence of numbers $\{y_n(x)\}$ converges for each x in I, we say that $\{y_n\}$ *converges to y pointwise on I* and denote this $\lim_{n\to\infty} y_n(x) = y(x)$ $(x$ in $I)$. Simply writing out the formal ϵ-definition gives us the equivalent statement: $\{y_n\}$ *converges to y pointwise on I* if for every x in I and every positive number ϵ, there is an integer N (depending on both x and ϵ) such that $|y_n(x) - y(x)| < \epsilon$ if $n \ge N$.

Now we forget temporarily the preceding definition and ask what kind of geometric interpretation should go with the idea of convergence of a sequence of functions. One reasonable idea is that the graphs of the functions y_n should approach the graph of the limit function y. This is a sensible idea, but it is not a consequence of the definition of pointwise convergence.

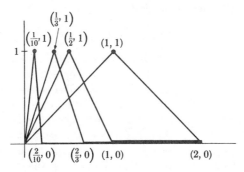

FIG. 6–1. Functions converging pointwise but not uniformly.

EXAMPLE 1. (cf. Fig. 6–1.) Let y_n $(n = 1, 2, \ldots)$ be the function on $[0, 2]$ whose graph consists of the line segments from $(0, 0)$ to $(1/n, 1)$, from $(1/n, 1)$ to $(2/n, 0)$, and from $(2/n, 0)$ to $(2, 0)$. This sequence converges to the function identically zero on $[0, 2]$ (Problem 2), even though each function is at some point one unit away from the x-axis.

We define next a type of convergence for sequences of functions which is a formalization of the statement that the graphs of the individual functions approach the graph of the limit function.

The sequence $\{y_n\}$ of functions on I *converges to y uniformly on I* if for every positive number ϵ, there is an integer N (which depends on ϵ but not on x) such that $|y_n(x) - y(x)| < \epsilon$ for all x in I, if $n \geq N$.

It is clear that uniform convergence is a stronger hypothesis than pointwise convergence. Some of the convenient properties of uniform convergence—and significant reasons for making such a definition—are given in the next two theorems.

THEOREM 4. *If $\{y_n\}$ is a sequence of continuous functions on $[a, b]$ which converges uniformly on $[a, b]$ to y, then y is continuous on $[a, b]$.*

This theorem is not true for pointwise convergence (Problem 4).

THEOREM 5. *If $\{y_n\}$ is a sequence of continuous functions on $[a, b]$ which converges uniformly on $[a, b]$ to y, then*

$$\lim_{n \to \infty} \int_a^b y_n(t)\, dt = \int_a^b y(t)\, dt.$$

This theorem also is false with the weaker hypothesis of pointwise convergence (Problem 3). The student is asked to supply a proof of Theorem 5 in Problem 6.

Now recall the connnection between series (infinite sums) and sequences. We say the series $\sum_{k=1}^{\infty} a_k$ *converges to a* (denoted $\sum_{k=1}^{\infty} a_k = a$) if the sequence $\{S_n\}$ converges to a, where

$$S_n = a_1 + a_2 + \cdots + a_n.$$

The sequence $\{S_n\}$ is called the *sequence of partial sums of $\sum_{k=1}^{\infty} a_k$*.

If $\{u_k\}$ is a sequence of functions defined on some interval I, we can form the series with these terms, $\sum_{k=1}^{\infty} u_k$. The convergence of a series of functions depends in the obvious way on the convergence of the sequence $\{S_n\}$ of functions which are the partial sums: $S_n(x) = u_1(x) + \cdots + u_n(x)$ for x in I. Thus we agree that $\sum_{k=1}^{\infty} u_k$ converges pointwise on I if the sequence $\{S_n\}$ converges pointwise on I, and $\sum_{k=1}^{\infty} u_k$ converges uniformly on I if $\{S_n\}$ converges uniformly on I. We will use the following comparison test.

THEOREM 6. *If for each k, u_k is a non-negative function on I, and $\sum_{k=1}^{\infty} u_k$ converges uniformly on I, and $|y_k(x)| \leq u_k(x)$ for each k and each x in I, then $\sum_{k=1}^{\infty} y_k$ converges uniformly on I.*

PROBLEMS

1. Show (given $\epsilon > 0$, find $\delta > 0$) that if f is continuous at a, and $f(a) = b$, and F is continuous at (a, b), and $g(t) = F\big(t, f(t)\big)$, then g is continuous at a.

2. (a) Show that the sequence of functions $\{y_n\}$ given in Example 1 converges pointwise to the function identically zero on $[0, 2]$; that is, give an integer N_x for each x in $[0, 2]$ such that $|y_n(x) - 0| < \epsilon$ if $n \geq N_x$. [*Hint:* it is sufficient to give an N_x such that $y_n(x) = 0$ if $n \geq N_x$.]

   (b) Show that if the functions of Example 1 are changed in any way on the intervals on which they are positive (but only there), the new sequence still converges to zero.

3. Modify the functions of Example 1 to obtain a sequence $\{y_n\}$ of functions such that $\lim_{n \to \infty} y_n(x) = 0$ (x in $[0, 2]$), and $\int_0^2 y_n(x)\, dx = 1$ for all n [cf. Problem 2(b)].

4. Let y_n be the sequence of functions on $[-1, 1]$ defined by

$$y_n(x) = \begin{cases} 0 & \text{if} \quad -1 \leq x \leq 0, \\[2mm] nx & \text{if} \quad 0 \leq x \leq \dfrac{1}{n}, \\[2mm] 1 & \text{if} \quad \dfrac{1}{n} \leq x \leq 1. \end{cases}$$

Graph y_1, y_2, y_3, and y_4. Show that $\{y_n\}$ converges on $[-1, 1]$. Graph the limit function. Does the sequence converge uniformly on $[-1, 1]$?

5. Give an example to show that the outside absolute value signs on the right side of (2) are necessary. [*Hint:* When is $\int_a^b |f(t)|\, dt$ negative?]

6. Show that if $\{y_n\}$ is a sequence of continuous functions on $[a, b]$ which converges uniformly on $[a, b]$ to y, then

$$\lim_{n \to \infty} \int_a^b y_n(t)\, dt = \int_a^b y(t)\, dt.$$

[*Hint:* Use (3) with f replaced by $y_n - y$. Compare this result with Problem 3.]

7. Let $\{S_n\}$ be any sequence. Find $a_1, a_2, \ldots$ such that $\{S_n\}$ is the sequence of partial sums for the series $\sum_{k=1}^{\infty} a_k$.

8. Show that if u_k is a continuous function on I for each k, and $\sum_{k=1}^{\infty} u_k$ converges uniformly on I to the function u, then u is continuous on I.

9. Show that if $|y_k(x)| \leq |x|^k/k!$ for all x in $[-1, 1]$ and all k, then $\sum_{k=1}^{\infty} y_k$ converges uniformly on $[-1, 1]$.

10. Let $y_n(x) = x^n$, $z_n(x) = x^n/n!$ Discuss the intervals on which the sequences $\{y_n\}$ and $\{z_n\}$ converge and the intervals on which the sequences converge uniformly.

6–2 Outline of the Picard method. We will prove in this section and the next the existence and uniqueness of a solution of the equation

$$y'(x) = F(x, y(x)), \qquad y(a) = b. \tag{1}$$

The equation (1) can be written in an equivalent integral form, as we show in Theorem 1, and it is this integral form which is principally used in the rest of the chapter.

THEOREM 1. *If F is continuous on the square* $S = \{(x, y): |x - a| \le h$ *and* $|y - b| \le h\}$ *and y is a continuous function on* $I = \{x:|x - a| \le h\}$ *whose graph is contained in S, then y is a solution of* (1) *on I if and only if y satisfies the following equation identically on I:*

$$y(x) = b + \int_a^x F(t, y(t)) \, dt. \tag{2}$$

Proof. Notice that the integral on the right side of (2) exists for all x in I, since the integrand is continuous on I (see Problem 1, Section 6–1). First assume that y satisfies (2) identically on I. Then in particular

$$y(a) = b + \int_a^a F(t, y(t)) \, dt = b, \tag{3}$$

so y satisfies the initial condition of (1). From the continuity of the integrand in (2) we can conclude that y is differentiable on I (Theorem 2, Section 6–1) and that

$$y'(x) = F(x, y(x)) \tag{4}$$

for each x in I. That is, (4) is an identity on I, and y satisfies (1) on I.

Now suppose y satisfies (1) on I. The integrals of equal functions are obviously equal, so we have

$$\int_a^x y'(t) \, dt = \int_a^x F(t, y(t)) \, dt \tag{5}$$

for all x in I. The left side of (5) can be integrated to obtain $y(x) - y(a) = y(x) - b$, so y satisfies (2) on I.

We can now outline the basic idea of the Picard method. Define a sequence of functions $\{y_n\}$ as follows:

$$
\begin{aligned}
y_0(x) &= b, \\
y_1(x) &= b + \int_a^x F(t, b) \, dt, \\
y_2(x) &= b + \int_a^x F(t, y_1(t)) \, dt, \\
&\ \ \vdots \\
y_{n+1}(x) &= b + \int_a^x F(t, y_n(t)) \, dt, \qquad \text{etc.}
\end{aligned}
\tag{6}
$$

Suppose there is a limit function y such that

$$\lim_{n \to \infty} y_{n+1}(x) = y(x) \tag{7}$$

and

$$\lim_{n \to \infty} \left[b + \int_a^x F(t, y_n(t)) \, dt \right] = b + \int_a^x F(t, y(t)) \, dt. \tag{8}$$

Since the left sides of (7) and (8) are equal by definition (6), it follows that the limit function y satisfies

$$y(x) = b + \int_a^x F(t, y(t)) \, dt$$

and hence is a solution of (2) and (1).

The following three questions must be resolved for this approach to work:

I. Does the scheme (6) actually define all the functions y_n on some fixed interval I_0 around a?

II. Does the sequence defined by (6) converge on I_0?

III. Does the sequence of functions $\int_a^x F(t, y_n(t)) \, dt$ converge on I_0 as required by (8)?

Our proof of the existence of a solution will consist in verifying conditions I, II, and III under appropriate assumptions on F.

EXAMPLE 1. Consider the equation $y'(x) = 2xy(x)$, $y(0) = 1$. The equivalent integral form is

$$y(x) = 1 + \int_0^x 2ty(t) \, dt.$$

We let $y_0(x) = 1$, $y_1(x) = 1 + \int_0^x 2t \, dt$, and in general

$$y_{n+1}(x) = 1 + \int_0^x 2ty_n(t) \, dt. \tag{9}$$

Here the function $F(x, y) = 2xy$ is continuous everywhere, so all the integrals in (9) exist for all x, and all the functions y_n are defined everywhere (Condition I). We get the following formulas:

$$y_1(x) = 1 + \int_0^x 2t1 \, dt = 1 + x^2,$$

$$y_2(x) = 1 + \int_0^x 2t(1 + t^2) \, dt = 1 + x^2 + \tfrac{1}{2}x^4,$$

$$y_n(x) = 1 + x^2 + \frac{1}{2}x^4 + \frac{1}{3 \cdot 2} x^6 + \cdots + \frac{1}{n!} x^{2n}$$

$$= \sum_{k=0}^n \frac{1}{k!} (x^2)^k. \tag{10}$$

The functions y_n form the sequence of partial sums of the series for e^{x^2}. This series converges everywhere, which means the sequence $\{y_n(x)\}$ converges to e^{x^2} for each x. Condition II is therefore satisfied on any interval around zero.

Condition III requires that

$$\lim_{n\to\infty} \int_0^x 2ty_n(t)\,dt = \int_0^x 2te^{t^2}\,dt = e^{x^2} - 1 \tag{11}$$

for all x in some interval around zero. By (10)

$$\int_0^x 2ty_n(t)\,dt = \int_0^x 2t \sum_{k=0}^n \frac{1}{k!} t^{2k}$$

$$= \sum_{k=0}^n \frac{2}{k!} \int_0^x t^{2k+1}\,dt$$

$$= \sum_{k=0}^n \frac{2}{k!} \frac{1}{2(k+1)} x^{2k+2}$$

$$= \sum_{k=0}^n \frac{1}{(k+1)!} (x^2)^{k+1}$$

$$= \sum_{k=1}^{n+1} \frac{1}{k!} (x^2)^k. \tag{12}$$

The functions (12) are the partial sums of the series for $e^{x^2} - 1$, and therefore converge to this function on every interval as required in (11). Condition III is satisfied and the limit e^{x^2} of the sequence $\{y_n(x)\}$ is a solution.

Problems

1. Write a first order differential equation with initial condition which is equivalent to

$$y(x) + 2\int_1^x y(t)\,dt = 1.$$

2. Write a second order differential equation with initial conditions which is equivalent to

$$y(x) + x\int_0^x y(t)\,dt = x^2 + 1.$$

3. Find the differentiable function y which satisfies

(a) $\int_0^x y(t)\,dt = x^2 - y(x)$,   (b) $xy(x) + \int_1^x y(t)\,dt = x^2.$

4. Compute the sequence $\{y_n\}$ defined by (6) for the equation $y' - y = 0$, $y(0) = 1$. Verify that conditions I, II, and III are satisfied for this sequence.

5. Compute the functions y_n defined by

$$\begin{cases} y_0(x) = 0, \\ y_{n+1}(x) = 1 + \int_0^x F\big(t, y_n(t)\big)\, dt, \end{cases}$$

where $F(x, y) = y$. Compare with Problem 4.

6. Find the sequence $\{y_n\}$ defined by (6) for the equation $y' = y - x$, $y(0) = 1$. Verify I, II, and III for any finite interval $[-r, r]$.

7. Use the Picard method, as in Problem 6, to solve $y' = x^2 - 1 - y$, $y(0) = 1$.

8. Show that III is not an automatic consequence of II. [*Hint:* Let $F(x, y) = y$ and use the sequence $\{y_n\}$ of Problem 3, Section 6–1.]

9. Let $\{y_n\}$ be a sequence of continuous functions on [0, 1] which converges *uniformly* on [0, 1] to the function y. Let $F(x, y) = xy$. Show that for x in [0, 1],

$$\lim_{n \to \infty} \int_0^x F\big(t, y_n(t)\big)\, dt = \int_0^x F\big(t, y(t)\big)\, dt.$$

(See Problem 6, Section 6–1.)

Answers

1. $y' + 2y = 0$, $y(1) = 1$
2. $y'' + xy' + 2y = 2$, $y(0) = 1$, $y'(0) = 0$
3. (a) $y(x) = 2x - 2 + 2e^{-x}$

   (b) $y(x) = \tfrac{2}{3}x + \dfrac{1}{3x^2}$

4. $y_n(x) = \displaystyle\sum_{k=0}^{n} x^k/k!$ 5. $y_n(x) = \displaystyle\sum_{k=0}^{n-1} x^k/k!$ $(n \geq 1)$

6. $y_n(x) = 1 + x - \dfrac{1}{(n+1)!} x^{n+1}$ 7. $y(x) = (1 - x)^2$

6–3 Proof of existence and uniqueness. Theorems 1, 2, and 3 of this section verify the conditions I, II, III of the preceding section and thus constitute a proof of the existence of a solution of

$$y'(x) = F\big(x, y(x)\big), \qquad y(a) = b, \tag{1}$$

or equivalently

$$y(x) = b + \int_a^x F\big(t, y(t)\big)\, dt. \tag{2}$$

The uniqueness of the solution is proved in Theorem 4, and the results of the chapter are summarized in Theorem 5 (Picard's theorem).

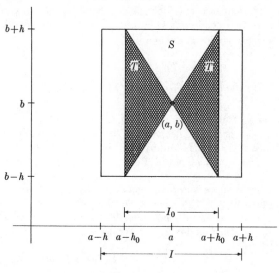

FIGURE 6–2

The notation introduced next and illustrated in Fig. 6–2 is used throughout this section.

Let $S = \{(x, y) : |x - a| \leq h \text{ and } |y - b| \leq h\}$.

Let $I = \{x : |x - a| \leq h\}$.

We will always assume that the function F in (1) and (2) is continuous on S, and it follows that F is bounded on S. Assume henceforth that $|F(x, y)| < M$ for (x, y) in S.

Draw the lines through (a, b) with slopes $\pm M$, and let T be the shaded double-triangle region between these lines and within S.

Let I_0 be the interval around a which is the projection of T on the x-axis. Let $h_0(h_0 \leq h)$ be the half-length of I_0, so that

$$I_0 = \{x : |x - a| \leq h_0\}$$

and

$$T = \{(x, y) : x \text{ in } I_0 \text{ and } |y - b| \leq M|x - a|\}. \tag{3}$$

We recall the inductive definition of the approximating sequence $\{y_n\}$.

$$
\begin{aligned}
y_0(x) &= b, \\
&\vdots \\
y_{n+1}(x) &= b + \int_a^x F(t, y_n(t))\, dt.
\end{aligned}
\tag{4}
$$

THEOREM 1 (Condition I). *If F is continuous on S, and $|F(x, y)| < M$ for (x, y) in S, and I_0, T are as indicated above, and $\{y_n\}$ is the sequence defined by (4), then each function y_n is defined on I_0, and has its graph in T.*

Proof. The proof is inductive. The function y_0 obviously has its graph over I_0 in T, so y_1 is well defined on I_0 by (4). Also,

$$|F(t, y_0(t))| = |F(t, b)| < M$$

for all t in I_0. For x in I_0,

$$\begin{aligned}
|y_1(x) - b| &= \left| \int_a^x F(t, y_0(t))\, dt \right| \\
&\leq \left| \int_a^x M\, dt \right| \\
&= M|x - a|.
\end{aligned} \tag{5}$$

The inequality (5) is just the condition (3) that $(x, y_1(x))$ be in T for x in I_0. Now suppose that above I_0 the graph of $y_n(n \geq 1)$ is in T. In particular, F is continuous on the graph of y_n and so y_{n+1} is defined for x in I_0. Also $|F(t, y_n(t))| < M$ for t in I_0, and hence

$$\begin{aligned}
|y_{n+1}(x) - b| &= \left| \int_a^x F(t, y_n(t))\, dt \right| \\
&\leq M|x - a|.
\end{aligned} \tag{6}$$

Thus y_{n+1} also has its graph in T by (3). This completes the proof by induction.

COROLLARY. *Under the hypotheses of Theorem 1,*

$$|y_m(x) - y_n(x)| \leq 2M|x - a| \tag{7}$$

for all m and n and all x in I_0.

Proof. Problem 1.

We have shown that the scheme (4) does define a sequence of functions on some fixed interval I_0 around a. We will need an additional assumption on F to show that the sequence converges.

DEFINITION. The function F satisfies a Lipschitz condition on S if there is a number A such that

$$|F(x, y_1) - F(x, y_2)| \leq A|y_1 - y_2| \tag{8}$$

for all points (x, y_1) and (x, y_2) in S.

The Lipschitz condition can be interpreted geometrically by saying that along any given vertical line (any fixed x), the function $F(x, y)$, considered as a function of the single variable y, nowhere increases more rapidly than the linear function Ay. This is true, for example, of the function $F(x, y) = x^2 + xy$ on any region on which x is bounded. On the other hand, the function $F(x, y) = 2\sqrt[3]{xy}$ (cf. Section 1–3) does not satisfy a Lipschitz condition in any region around $(1, 0)$. Here Formula (8) would require, in particular ($x = 1$), that

$$|F(1, y) - F(1, 0)| = |2\sqrt[3]{y}| \leq A|y - 0| = A|y|$$

for all y in some interval around zero. This would say that for some constant A and all small values of y,

$$\left| \frac{2\sqrt[3]{y}}{y} \right| = 2y^{-2/3} \leq A,$$

which is impossible.

THEOREM 2 (Condition II). *If F is continuous on S, and $|F(x, y)| < M$ for (x, y) in S and F satisfies the Lipschitz condition (8), then the sequence $\{y_n\}$ defined by (4) converges uniformly on I_0.*

Proof. We will consider instead of the sequence $\{y_n\}$, the series whose nth partial sum, for each n, is y_n. The uniform convergence of the sequence $\{y_n\}$ is by definition equivalent to the uniform convergence of this series. We show that the terms of the series are smaller than the corresponding terms of a uniformly convergent power series, and thus complete the proof.

Recall the identity (cf. Problem 7, Section 6–1)

$$y_n(x) = y_0(x) + [y_1(x) - y_0(x)] + \cdots + [y_n(x) - y_{n-1}(x)]. \quad (9)$$

The sequence $\{y_n\}$ is the sequence of partial sums of the series

$$y_0(x) + \sum_{n=0}^{\infty} [y_{n+1}(x) - y_n(x)]. \quad (10)$$

We will show inductively that

$$|y_n(x) - y_{n-1}(x)| \leq \frac{2MA^{n-1}|x - a|^n}{n!} \quad (11)$$

for all x in I_0. The terms on the right of (11) are the terms of the series

$$\frac{2M}{A} \sum_{n=0}^{\infty} \frac{A^n|x - a|^n}{n!} = \frac{2M}{A} e^{A|x-a|}. \quad (12)$$

The series (12) converges uniformly on every finite interval—in particular on I_0—and hence the inequality (11) shows that (10) converges uniformly on I_0.

To prove (11), first consider $n = 2$:

$$
\begin{aligned}
|y_2(x) - y_1(x)| &= \left| \int_a^x [F(t, y_1(t)) - F(t, y_0(t))]\, dt \right| \\
&\leq \left| \int_a^x A|y_1(t) - y_0(t)|\, dt \right| \\
&\leq \left| \int_a^x A2M|t - a|\, dt \right| \\
&= \frac{2MA|x - a|^2}{2!}.
\end{aligned}
\tag{13}
$$

In (13) we used the Lipschitz condition (8) to show that

$$|F(t, y_1(t)) - F(t, y_0(t))| \leq A|y_1(t) - y_0(t)|$$

and (7) to see that

$$|y_1(t) - y_0(t)| \leq 2M|t - a|.$$

Now suppose that (11) holds for some $n \geq 2$. Then

$$
\begin{aligned}
|y_{n+1}(x) - y_n(x)| &= \left| \int_a^x [F(t, y_n(t)) - F(t, y_{n-1}(t))]\, dt \right| \\
&\leq \left| \int_a^x A|y_n(t) - y_{n-1}(t)|\, dt \right| \\
&\leq \left| \int_a^x A\, \frac{2MA^{n-1}|t - a|^n}{n!}\, dt \right| \\
&= \frac{2MA^n|x - a|^{n+1}}{(n + 1)!}.
\end{aligned}
\tag{14}
$$

Thus if (11) holds for n, it holds for $n + 1$ and, by induction, the inequality holds for all $n \geq 2$. Since the series (12) converges uniformly on I_0, the smaller series (10) converges uniformly on I_0, which is the same as saying the approximating sequence $\{y_n\}$ converges uniformly on I_0.

COROLLARY. *If y is the limit on I_0 of $\{y_n\}$ then y is continuous on I_0 and has its graph in T.*

Proof. Problem 2.

THEOREM 3 (Condition III). *If F is continuous on S and $|F(x, y)| < M$ for (x, y) in S, and F satisfies the Lipschitz condition* (8) *on S, and y is the limit of $\{y_n\}$ on I_0, then*

$$\lim_{n \to \infty} \int_a^x F(t, y_n(t))\, dt = \int_a^x F(t, y(t))\, dt \tag{15}$$

for all x in I_0.

Proof. The equality above can be written

$$\lim_{n \to \infty} \int_a^x [F(t, y_n(t)) - F(t, y(t))]\, dt = 0. \tag{16}$$

By (8) we have

$$\left| \int_a^x |F(t, y_n(t)) - F(t, y(t))|\, dt \right|$$

$$\leq \left| \int_a^x A|y_n(t) - y(t)|\, dt \right|$$

$$\leq A|x - a| \max_{t \text{ in } I_0} |y_n(t) - y(t)|$$

$$\leq Ah_0 \max_{t \text{ in } I_0} |y_n(t) - y(t)|. \tag{17}$$

Since y_n converges *uniformly* on I_0 to y, for any $\epsilon > 0$ we can find N such that

$$\max_{t \text{ in } I_0} |y_n(t) - y(t)| < \frac{\epsilon}{Ah_0}$$

for all $n \geq N$. Hence for $n \geq N$,

$$\left| \int_a^x [F(t, y_n(t)) - F(t, y(t))]\, dt \right| < \epsilon$$

which verifies (16) and completes the proof.

Theorems 1, 2, and 3, together with our observations of Section 6–2, complete the proof of the existence of a solution. We proceed to the proof that the solution is unique.

The following lemma shows that *any* solution of (1) or (2) has its graph in T. We will need this in Theorem 4 (proof of uniqueness) so we can apply the Lipschitz condition to any solution g, whether it arises as the limit of a Picard sequence or not.

LEMMA 1. *If F is continuous on S, and $|F(x, y)| < M$ for (x, y) in S, and g is a solution of* (1) *or* (2) *on I_0, then the graph of g is in T.*

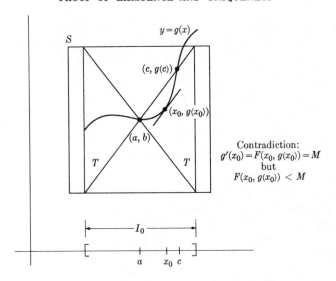

FIG. 6–3. The graph of a solution lies in T.

Proof. If g is a solution on I_0, then $g'(a) = F(a, b)$ and hence $|g'(a)| < M$. It follows (Problem 4) that the graph of g starts into T; that is, that

$$|g(x) - b| < M|x - a|$$

for all x in some interval $(a - \delta, a + \delta)$. If the graph of g leaves T, say to the right of a, then by the continuity of g (Problem 5) there will be a smallest number $c > a$ where the graph crosses one of the diagonal lines (Fig. 6–3). That is, c will be the first number to the right of a such that

$$|g(c) - b| = M|c - a|. \tag{18}$$

By the Mean Value Theorem, there is a number x_0 in (a, c) such that

$$|g(c) - g(a)| = |g'(x_0)||c - a|. \tag{19}$$

By the definition of c, $(x_0, g(x_0))$ is in T, and hence

$$|g'(x_0)| = |F(x_0, g(x_0))| < M.$$

From (19), we conclude that

$$|g(c) - b| < M|c - a|$$

which contradicts (18). Therefore the graph of g cannot leave T.

THEOREM 4 (Uniqueness of the solution). *If F is continuous on S and* $|F(x, y)| < M$ *for* (x, y) *in S, and F satisfies the Lipschitz condition* (8) *on S and y, g are any two solutions of* (2) *over the interval* I_0, *then* $y(x) = g(x)$ *for all x in* I_0.

Proof. The hypothesis is that $y(a) = g(a) = b$, and

$$y'(x) = F(x, y(x)),$$
$$g'(x) = F(x, g(x)),$$

for all x in I_0. We show that if y and g have the same value at any point x_0 if I_0, then $y(x) = g(x)$ on the interval $[x_0 - 1/(2A), x_0 + 1/(2A)]$. By starting at $x_0 = a$, we get $y(x) = g(x)$ on $[a - 1/(2A), a + 1/(2A)]$. Then taking $x_0 = a + 1/(2A)$, we get $y(x) = g(x)$ on $[a, a + 2/(2A)]$. With a finite number of repetitions of this argument, we show that $y(x) = g(x)$ on all of I_0.

Assume, therefore, that $y(x_0) = g(x_0)$ for some x_0 in I_0, and let $h(x) = y(x) - g(x)$. We have $h(x_0) = 0$, and we must show that $h(x) = 0$ on $[x_0 - 1/(2A), x_0 + 1/(2A)]$. By Lemma 1, and (8), we have

$$\begin{aligned}
|h'(x)| &= |y'(x) - g'(x)| \\
&= |F(x, y(x)) - F(x, g(x))| \\
&\leq A|y(x) - g(x)| \\
&= A|h(x)|.
\end{aligned} \tag{20}$$

Since h is continuous on $[x_0 - 1/(2A), x_0 + 1/(2A)]$, there is a point of this interval where $|h(x)|$ assumes its maximum value—say $|h(x_1)|$ is this maximum value. By the Law of the Mean and the fact that $h(x_0) = 0$, we have

$$\begin{aligned}
|h(x_1)| &= |h(x_1) - h(x_0)| \\
&= |h'(\xi)||x_1 - x_0|,
\end{aligned} \tag{21}$$

for some number ξ between x_1 and x_0. From (20) and (21) we get

$$\begin{aligned}
|h(x_1)| &= |h'(\xi)||x_1 - x_0| \\
&\leq A|h(\xi)||x_1 - x_0| \\
&\leq A|h(x_1)| \frac{1}{2A} \\
&= \tfrac{1}{2}|h(x_1)|.
\end{aligned}$$

The inequality $|h(x_1)| \leq \tfrac{1}{2}|h(x_1)|$ implies that the maximum value $|h(x_1)| = 0$, and hence that $h(x) \equiv 0$ on $[x_0 - 1/(2A), x_0 + 1/(2A)]$.

We recapitulate the results of the preceding theorems in Theorem 5.

THEOREM 5 (Picard's Theorem). *If F is continuous on a square S with center (a, b), and F satisfies (8) on S, then there is an interval I_0 around a such that there is, on I_0, exactly one solution of (1).*

COROLLARY. *If F and F_y are continuous on a square S with center (a, b), then on some interval I_0 around a there is one and only one solution of (1).*

Proof. The conclusion of the corollary is merely a paraphrase of the conclusion of the Theorem. To prove the corollary is clearly sufficient to show that the continuity of F_y implies that the Lipschitz condition (8) holds. The student is asked to carry out the details in Problem 7.

PROBLEMS

1. Prove the corollary to Theorem 1.
2. Prove the corollary to Theorem 2. [*Hint:* Give a reason why each y_n is continuous and cite the appropriate theorem from Section 6–1.]
3. Prove that the integral on the right side of (15) exists.
4. Assume the hypotheses of Lemma 1. Let $\epsilon = M - |F(a, b)| > 0$. There is (why?) a positive number δ such that

$$\left| \frac{g(x) - b}{x - a} - F(a, b) \right| < M - |F(a, b)|$$

if $0 < |x - a| < \delta$. Show that this implies, as stated in the proof of Lemma 1, that $|g(x) - b| \leq M|x - a|$ if $|x - a| < \delta$. [*Hint:* $|A - B| < C$ is equivalent to $B - C < A < B + C$, $-M \leq F(a, b) - \epsilon$, and $-D \leq A \leq D$ is equivalent to $|A| \leq D$.]
5. In the proof of Lemma 1, it is stated that g is continuous. Why is g continuous on I_0?
6. Show that $y(x) = x^2$ is not a solution on [0, 1] of any equation $y' = F(x, y)$ with $|F(x, y)| < 1$ for $0 \leq x \leq 1$, $0 \leq y \leq 1$.
7. Prove the corollary of Theorem 5. [*Hint:* By its continuity, F_y must be bounded on S; say $|F_y(x, y)| < A$ for (x, y) in S. Use the Mean Value Theorem for the function $f(y) = F(x, y)$, x any fixed number in I (Fig. 6–2), to verify that (8) holds on S.]
8. Show by the methods of this section that the equation $y' = 1/(1 + y^2)$, $y(0) = 1$, has a solution on every interval $I_0 = [-h_0, h_0]$, and hence a solution on the whole line.
9. Show that $[-\frac{1}{2}, \frac{1}{2}]$ is the biggest interval around zero on which the methods of this section will guarantee a solution to $y' = 1 + y^2$, $y(0) = 0$. [*Hint:* Show that the optimum square S to start with has sides of length 2 (that is, $h = 1$).]

6–4 Approximations to solutions. In many of the applications of differential equations, the primary interest is in finding numerical values of a given solution. Most numerical statements are statements of approximation (e.g., $\pi = 3.1416$, $\sqrt{2} = 1.414$, $\sin \pi/4 = .707$), and *all* of the measured quantities which the scientist and engineer are concerned with are approximations. For most numerical applications, therefore, there is no useful distinction between an approximate solution and an exact solution, provided the approximation is sufficiently good. It should be noted also that even when a differential equation can be solved exactly, the numerical values of the solution may not be readily accessible. For example, the equation

$$y^3 + y = x \tag{1}$$

characterizes the solution of

$$y' = \frac{1}{3y^2 + 1}, \qquad y(0) = 0, \tag{2}$$

but the values of this solution y cannot be obtained from (1) without further computation. Similarly, the solution of the linear equation

$$y' = 1 - 2xy, \qquad y(0) = 0 \tag{3}$$

can be written

$$y = e^{-x^2} \int_0^x e^{t^2}\, dt, \tag{4}$$

but the Formula (4) by itself does not immediately yield numerical values of y.

The Picard proof not only shows that there is a solution of the equation

$$y' = F(x, y), \qquad y(a) = b, \tag{5}$$

but gives a theoretical method of calculating approximations to the solution. Usually, however, the successive integrals which define the approximating functions become unmanageable, and the Picard method is not an effective way of obtaining approximations. The first proof of an existence theorem for (5), due primarily to Cauchy, is not so elementary as Picard's proof, but it does provide an effective way of finding approximations to the solution. Cauchy's method of proof, like Picard's, consists in defining a sequence of approximate solutions, showing that this sequence converges on some interval, and proving that the limit function must be a solution. The approximating functions of the Cauchy method are polygonal curves obtained by following the direction field along short segments. The calculations involved in finding these approximations are purely arithmetic, and therefore well adapted to machine computation.

To construct a polygonal approximation to the solution of (5) on an interval $[a, c]$, we proceed as follows. Let $\{x_0, x_1, \ldots, x_n\}$ be a partition of $[a, c]$; that is, let $a = x_0 < x_1 < x_2 < \cdots < x_n = c$. Compute $F(a, b)$, and follow the line through (a, b) with slope $F(a, b)$ until it intersects the line $x = x_1$, say at the point (x_1, y_1). Then follow the line through (x_1, y_1) with slope $F(x_1, y_1)$ until it intersects the line $x = x_2$. Continuing in this way, we obtain a sequence of points $(a, b), (x_1, y_1), \ldots, (x_n, y_n)$; the broken line path joining these points is the *polygonal approximation for the partition* $\{x_0, x_1, \ldots, x_n\}$. If we let $y_0 = b$, then we can write

$$y_{k+1} = y_k + F(x_k, y_k)(x_{k+1} - x_k), \qquad k = 0, 1, \ldots, n - 1.$$

Since $y_n = y_0 + [y_1 - y_0] + \cdots + [y_n - y_{n-1}]$, we have also

$$y_n = y_0 + \sum_{k=0}^{n-1} F(x_k, y_k)(x_{k+1} - x_k). \tag{6}$$

A polygonal approximation to the solution on an interval $[d, a]$ to the left of the initial point can be constructed in a similar way. If $d = x_n < x_{n-1} < \cdots < x_1 < x_0 = a$, then the vertices (x_k, y_k) of the approximation are determined by

$$\frac{y_k - y_{k+1}}{x_k - x_{k+1}} = F(x_k, y_k), \qquad k = 0, 1, \ldots, n - 1,$$

and again we have

$$y_{k+1} = y_k + F(x_k, y_k)(x_{k+1} - x_k).$$

In general, one gets better approximations by taking finer partitions, and for any partition the approximation is likely to be best close to a.

For the sake of completeness we will state without proof* a theorem which indicates how polygonal approximations can be used to show the existence of a solution of (5).

THEOREM 1. *Assume F satisfies the hypotheses of Section 6–3. Let p_n be a sequence of partitions of $I_0 = [a - h_0, a + h_0]$ such that if $\|p_n\|$ is the maximum distance between points of p_n, then $\lim_{n \to \infty} \|p_n\| = 0$. If y_{p_n} is the polygonal approximation for p_n, then the sequence $\{y_{p_n}\}$ converges uniformly on I_0, and the limit function is a solution of (5).*

The following example shows a tabulation of the computations involved in finding a polygonal approximation. We will find two approximations

---

* For a proof and further discussion see, e.g., Hurewicz, *Lectures on Ordinary Differential Equations* (New York: John Wiley and Sons, Inc., 1958), pp. 1–12.

to the solution of (3) on the interval [0, 2], using two different partitions of the interval.

EXAMPLE 1. $y' = 1 - 2xy$, $y(0) = 0$.

TABLE 1

Partition {0, 0.5, 1.0, 1.5, 2.0}

x_k	0	0.5	1.0	1.5	2.0
y_k	0	0.5	0.75	0.5	0.25
$F(x_k, y_k) = 1 - 2x_k y_k$	1	0.5	−0.5	−0.5	
$\Delta y_k = \frac{1}{2}(1 - 2x_k y_k)$	0.5	0.25	−0.25	−0.25	
$y_k + \Delta y_k = y_{k+1}$	0.5	0.75	0.5	0.25	

TABLE 2

Partition {0, 0.25, 0.5, 0.75, 1.0, 1.25, 1.5, 1.75, 2.0}

x_k	0	0.25	0.50	0.75	
y_k	0	0.25	0.4688	0.6016	
$F(x_k, y_k) = 1 - 2x_k y_k$	1	0.875	0.5312	0.0976	
$\Delta y_k = \frac{1}{4}(1 - 2x_k y_k)$	0.25	0.2188	0.1328	0.0244	
$y_k + \Delta y_k = y_{k+1}$	0.25	0.4688	0.6016	0.6260	
x_k	1.0	1.25	1.5	1.75	2.0
y_k	0.6260	0.5630	0.4611	0.3653	0.2956
$F(x_k, y_k) = 1 - 2x_k y_k$	−0.2520	−0.4075	−0.3833	−0.2789	
$\Delta y_k = \frac{1}{4}(1 - 2x_k y_k)$	−0.0630	−0.1019	−0.0958	−0.0697	
$y_k + \Delta y_k = y_{k+1}$	0.5630	0.4611	0.3653	0.2956	

The graphs of these two approximations are shown in Fig. 6–4. Although one cannot conclude directly from Theorem 1 that the polygonal approximations will converge on an interval as large as [0, 2], this can be shown by a more careful examination of the particular function F in this example.

It is not easy to make a precise estimate of the accuracy of an approximation. Suppose one requires, for example, accuracy to two decimal places. The usual procedure is to compute approximations for finer and finer partitions until two successive approximations agree to two decimal places. Since the error from rounding off decimals may well accumulate from step to step in computing an approximation, it is usually necessary to carry several more decimal places in the calculations than the accuracy desired.

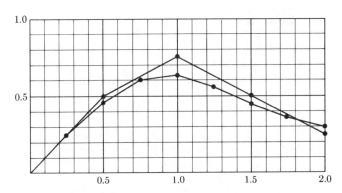

FIG. 6–4. Polygonal approximations for $y' = 1 - 2xy$, $y(0) = 0$.

PROBLEMS

1. Find a polygonal approximation on $[0, \frac{1}{2}]$ of the solution to $y' = 1 - 2xy$, $y(0) = 0$. Use first the partition $\{0, 0.1, 0.2, 0.3, 0.4, 0.5\}$ and then the partition $\{0, 0.05, 0.1, 0.15, \ldots, 0.45, 0.5\}$.

2. Find polygonal approximations to the solution of $y' = x + y^2$, $y(0) = 0$, on the interval $[0, 1]$. Partition the interval into subintervals of length 0.2 and then into subintervals of length 0.1. Plot the two approximations on the same graph.

3. Find and graph the polygonal approximation to the solution of $y' = 1/(x + y)$, $y(1) = 1$.

(a) Use the partition $\{1, 1.1, 1.2, 1.3, 1.4, 1.5\}$ of $[1, 1.5]$.

(b) Use the partition $\{0.5, 0.6, 0.7, 0.8, 0.9, 1.0\}$ of $[0.5, 1]$.

4. Let y_p be the polygonal approximation to the solution of $y' = f(x)$, $y(a) = 0$, for the partition $p = \{a = x_0, x_1, \ldots, x_n = c\}$ of $[a, c]$. Show that $y_p(c)(= y_n)$ is a Riemann sum approximation to $\int_a^c f(x)\, dx$ [cf. Formula (6)].

5. Compute an approximate value for $e^{0.1}$ by finding the polygonal approximation to the solution of $y' = y$, $y(0) = 1$ for the partition $\{0, 0.01, 0.02, \ldots, 0.09, 0.1\}$ of $[0, 0.1]$. Carry five decimal places. (To four decimal places, $e^{0.1} = 1.1052$.)

6. Find the polygonal approximation on $[1, 2]$ of the solution of $y' = 2 - y/x$, $y(1) = 2$. Use

(a) the partition $\{1, 1.2, 1.4, \ldots, 1.8, 2.0\}$

(b) the partition $\{1, 1.1, 1.2, \ldots, 1.9, 2.0\}$.

Solve the equation and plot the solution and the two approximations on the same graph.

7. The solution of the problem of Example 1 is analytic [cf. Formula (4)], and the *Picard* approximations are the partial sums of the power series for the solution. Find the Picard approximations y_0, y_1, y_2, y_3, y_4, and evaluate $y_4(0.5)$. Show that $y_4(0.5)$ is accurate to within 0.000033. [*Hint:* The series is alternating, so the error is less than the first term omitted: $16(\frac{1}{2})^9/3 \cdot 5 \cdot 7 \cdot 9$).] Also find

$y_4(1)$, $y_4(2)$, estimate the error and compare these values with the polygonal approximations of Table 2.

8. Use the differential equation (2) and the partition $\{0, 0.1, 0.2, \ldots, 0.9, 1.0\}$ to obtain an approximate graph over $[0,1]$ for the function y defined by (1). Let $x = 1$ and solve (1) for y to three decimal places by trial and error.

ANSWERS

1. 1st partition y_k: 0, 0.1, 0.198, 0.2901, 0.3727, 0.4429
   2nd partition y_k: 0, 0.05, 0.0998, 0.1488, 0.1966, 0.2427, 0.2866, 0.3280, 0.3665, 0.4018, 0.4337
2. 1st partition y_k: 0, 0, 0.04, 0.1203, 0.2432, 0.4150
   2nd partition y_k: 0, 0, 0.01, 0.0300, 0.0601, 0.1005, 0.1515, 0.2138, 0.2884, 0.3767, 0.4809
3. (a) 1, 1.05, 1.097, 1.141, 1.182, 1.221
   (b) 1, 0.950, 0.896, 0.837, 0.772, 0.699
5. 1.1046
6. (a) 2, 2, 2.07, 2.17, 2.30, 2.44
   (b) 2, 2, 2.02, 2.05, 2.09, 2.14, 2.20, 2.26, 2.33, 2.40, 2.47
   Solution: $y = x + 1/x$
7. $y_4(0.5) = 0.42441$, max error 0.000033
   $y_4(1) = 0.525$, max error 0.017
   $y_4(2) = -4.6$, max error 8.7
8. 0, 0.1, 0.197, 0.287, 0.367, 0.438, 0.501, 0.558, 0.610, 0.657, 0.701;
   $y(1) = 0.682$ by trial

CHAPTER 7

SYSTEMS OF EQUATIONS

7–1 Geometric interpretation of a system. In this chapter we will treat systems of several simultaneous differential equations in several unknown functions. We first consider the case in which there are two unknown functions, y and z, and try to interpret the situation geometrically. The systems we want to consider are those of the form

$$
\begin{aligned}
y' &= F(x, y, z) \qquad \left(y' = \frac{dy}{dx}\right), \\[2mm]
z' &= G(x, y, z) \qquad \left(z' = \frac{dz}{dx}\right).
\end{aligned}
\tag{1}
$$

A *solution* of (1) is a pair of functions, (f, g), which satisfy (1) identically on some interval. That is, (f, g) is a solution of (1) on the interval I if for all x in I,

$$
\begin{aligned}
f'(x) &= F\big(x, f(x), g(x)\big), \\[2mm]
g'(x) &= G\big(x, f(x), g(x)\big).
\end{aligned}
\tag{2}
$$

It is convenient to indicate solutions of a system with a pair of equations; thus we write

$$
\begin{aligned}
y &= f(x), \\
z &= g(x)
\end{aligned}
\tag{3}
$$

to indicate that (f, g) is a solution of (1).

Now let us ask whether we can reasonably expect the system (1) to have solutions. First consider the geometric interpretation of the equations (3). The graph of $y = f(x)$ in three dimensions is a cylindrical surface consisting of the points covered by a line parallel to the z-axis moving along the curve $y = f(x)$ in the xy-plane (see Fig. 7–1). Similarly, the graph of the equation $z = g(x)$ is a cylinder consisting of lines parallel to the y-axis. The graph of the pair of equations (3) is the curve which is the intersection of these two surfaces. Hence we can picture a solution of the system (1) as being a curve in space.

Now let us ask what the system (1) demands of a curve in order that it be a solution. Recall that the line tangent to the curve (3) at a point

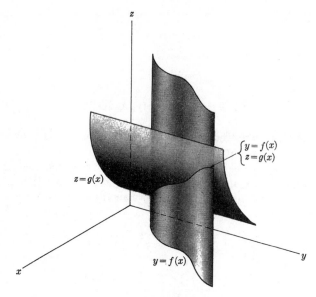

$$\begin{cases} y = f(x) \\ z = g(x) \end{cases}$$

$z = g(x)$

$y = f(x)$

FIGURE 7–1

$(x, f(x), g(x))$ has direction numbers $(1, f'(x), g'(x))$. Comparison with (2) and (1) shows that in effect the system (1) prescribes at each point (x, y, z) a tangent line for a solution curve; namely, the line through (x, y, z) with direction numbers $(1, F(x, y, z), G(x, y, z))$. Thus the system (1) describes a tangent field in space, just as the single equation $y' = F(x, y)$ describes a tangent field in the plane. It is true, as one would expect, that if F and G are continuous, then there is a solution curve for (1) through each point in space. In other words, starting at any point the tangent field described by (1) "directs" through space a curve which represents a solution. We will prove an existence theorem to this effect in Section 7–4 under an additional assumption (a Lipschitz condition) which allows us also to prove that the solution curve through any point is unique.

EXAMPLE 1. $y' = y/x, \qquad z' = x - yz.$

The first equation does not involve z and can readily be solved by separating variables. Accordingly, we know that for y and z to satisfy the system, we must in particular have $y = c_1 x$. Therefore z must satisfy

$$z' = x - c_1 xz,$$
$$z' + c_1 xz = x.$$

The function $e^{1/2(c_1 x^2)}$ is an integrating factor for this linear equation, so z is given by

$$ze^{1/2(c_1 x^2)} = \int x e^{1/2(c_1 x^2)}\, dx + c_2,$$

$$z = \frac{1}{c_1} + c_2 e^{-1/2(c_1 x^2)} \qquad (\text{if } c_1 \neq 0).$$

If $c_1 = 0$, so that $y = 0$, then $z' = x$, and $z = \frac{1}{2}x^2 + c_2$.

EXAMPLE 2. $y' = y - z, \qquad z' = y - z.$

Here both unknowns occur in both equations, so we cannot solve either equation by itself. However, $y' = z'$ if (y, z) is a solution, so we must have $y - z = c_1$ for any solution. Therefore y and z must satisfy

$$y' = c_1, \qquad\qquad z' = c_1,$$
$$y = c_1 x + c_2, \qquad z = c_1 x + c_3.$$

The argument above shows that the last equations are necessary for a solution, but not that they are sufficient. If we substitute the functions given above in either equation of the system, we get

$$c_1 = c_1 x + c_2 - c_1 x - c_3 = c_2 - c_3.$$

Therefore c_1, c_2, and c_3 are not independent, and the solutions of the system are given, for example, by

$$y = c_1 x + c_2, \qquad z = c_1 x + c_2 - c_1.$$

EXAMPLE 3. $y' = z + x, \qquad z' = 2z - y + x^2.$

Here we can use the technique of solving one equation for one unknown and substituting in the other. From the first equation, we have

$$z = y' - x, \tag{4}$$

and hence

$$z' = y'' - 1.$$

Substitution of these values for z and z' in the second equation of the system gives

$$y'' - 1 = 2(y' - x) - y + x^2,$$
$$y'' - 2y' + y = x^2 - 2x + 1. \tag{5}$$

The solutions of (5) are

$$y = c_1 e^x + c_2 x e^x + x^2 + 2x + 3. \tag{6}$$

From (6) and (4) it follows that if (y, z) is a solution of the system, then we must have

$$\begin{aligned} y &= c_1 e^x + c_2 x e^x + x^2 + 2x + 3, \\ z &= c_1 e^x + c_2(x+1)e^x + x + 2. \end{aligned} \tag{7}$$

Conversely, it is easy to show (Problem 8) that if y is a solution of (5) and z satisfies (4), then (y, z) is a solution of the system. That is, *all* pairs (y, z) given by (7) are solutions of the system.

Let us make use of the example above to note the connection between the existence theorem for second order equations and that for first order systems. Since z determines y' and *vice versa* by (4), there is a unique solution of (5) for any initial conditions $y(a) = b_0$, $y'(a) = b_1$ if and only if there is a unique solution of the system for any initial conditions $y(a) = b_0$, $z(a) = b_1$. We will use this sort of argument in Section 7–5 to prove the existence theorem for a single nth order equation, after having proved an existence theorem for first order systems.

PROBLEMS

1. Show that each line in the tangent field of the system $y' = y/x$, $z' = z/x$ points toward the origin. Describe the solution curves.

2. Describe the tangent field and solution curves of the system $y' = -x/y$, $z' = 0$.

3. Find the lines through the origin $(y = ax, z = bx)$ which are solution curves of the system $y' = xy/z^2$, $z' = yz/2x^2$.

4. Find the curves of the form $y = ax^2$, $z = bx$ which are solutions of the system $y' = (y + z^2)/x$, $z' = (3z^2 - 2xz)/y$.

5. Solve the system, and check your answer. $y' = z + y/x$, $z' = (x + z)/x$.

6. Solve the system, and check your answer. $y' = y + z$, $z' = y + z$. Give an example to show that the following is false: (y, z) is a solution if and only if $y - z = c$.

7. Solve the system and check your answer. $y' = 3x^2 + y - z$, $z' = y - z$. [*Hint:* Subtract the equations.]

8. Show that if y is a solution of (5), and z is given by (4), then (y, z) is a solution of the system of Example 3.

Solve the following systems.

9. $y' = z$,
    $z' = y$

10. $y' = y - z$,
     $z' = x - 2y$

11. $y' = z$,
     $z' = 2e^x - y$

12. $y' = y + z + x$,
     $z' = 4y + z + x + 4e^x$

13. $y' = x^2 + \frac{1}{3}z$,
     $z' = 3x^2 + 6y + z$

Answers

1. All lines through the origin except those in the yz-plane

2. The solutions are the circles parallel to the xy-plane with centers on the z-axis.

3. $y = 2x,\ z = x;\ y = 2x,\ z = -x$

4. $y = x^2\ \ z = x;\ y = 4x^2,\ z = 2x$

5. $y = x^2\,(\ln|c_2x| - 1) + c_1x,\ z = x\ln|c_2x|$

6. $y = c_2e^{2x} + \frac{1}{2}c_1,\ z = c_2e^{2x} - \frac{1}{2}c_1$

7. $y = x^3 + \frac{1}{4}x^4 + c_1x + c_2,\ z = \frac{1}{4}x^4 + c_1x + c_2 - c_1$

9. $y = c_1e^x + c_2e^{-x},\ z = c_1e^x - c_2e^{-x}$

10. $y = c_1e^{-x} + c_2e^{2x} + \frac{1}{2}x - \frac{1}{4},\ z = 2c_1e^{-x} - c_2e^{2x} + \frac{1}{2}x - \frac{3}{4}$

11. $y = c_1\cos x + c_2\sin x + e^x,\ z = -c_1\sin x + c_2\cos x + e^x$

12. $y = c_1e^{-x} + c_2e^{3x} - \frac{1}{3} - e^x,\ z = -2c_1e^{-x} + 2c_2e^{3x} + \frac{1}{3} - x$

13. $y = c_1e^{2x} + c_2e^{-x} - x + \frac{1}{2},\ z = 6c_1e^{2x} - 3c_2e^{-x} - 3 - 3x^2$

7–2 Other interpretations of a system. A system of two first order equations has interpretations other than the geometric one given in the preceding section. To see how such a system might arise in mechanics, consider a particle moving in the xy-plane and write $x = x(t)$, $y = y(t)$ for the coordinates of the particle at time t. Suppose that we know the velocity (speed and direction) that the particle must have if it passes through the point (x, y) at the time t. That is, we assume that we know the x-component dx/dt and the y-component dy/dt of the velocity in terms of x, y, and t. This gives us equations of the form

$$\frac{dx}{dt} = F(t, x, y),$$
$$\frac{dy}{dt} = G(t, x, y). \tag{1}$$

This system is identical to system (1) of the preceding section, except we have relabeled the independent variable t, and the unknown functions x and y.

Although the situation outlined above does occur, it is not typical in problems of motion. One is more likely to know the forces on a particle, in terms of position, time, and velocity, than to know the velocity in terms of position and time. In this case, the motion is described by the system of two second order equations obtained by equating $m(d^2x/dt^2)$ and $m(d^2y/dt^2)$ to the horizontal and vertical forces.

Another interpretation of (1) results if we regard t simply as a parameter, and a solution

$$x = f(t), \qquad y = g(t)$$

of (1) as being the parametric equations of a plane curve. For example,

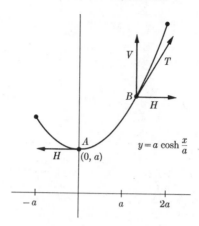

$$y = a \cosh \frac{x}{a}$$

FIG. 7-2. The catenary.

in setting up the differential equation of a curve it may be easier to determine the components dx/dt and dy/dt of the tangent vector in terms of a parameter t than it is to find dy/dx in terms of x and y. The following example illustrates this procedure of describing a plane curve by a pair of differential equations involving a parameter.

EXAMPLE 1. Find the equation of the curve (called a *catenary*) formed by a flexible cable suspended from two points (Fig. 7-2).

Let the linear density of the cable be ω, and let the tension in the cable at the low point (A) be H. Define a by the equation $H = a\omega$. It is convenient to introduce coordinates so that the coordinates of the point A are $(0, a)$. We will find the coordinates (x, y) of any point B on the curve in terms of the length s of the arc from A to B.

Let T be the tension in the cable at B. The arc from A to B is in equilibrium, so the horizontal component of T must equal the horizontal component H of the tension at A. Also, the vertical component V of T must equal the weight ωs of the arc AB. Since dx/ds and dy/ds are the components of a unit tangent vector at B, it follows that $T\,(dx/ds) = H$, and $T\,(dy/ds) = V$. We have the following formulas:

$$\frac{dx}{ds} = \frac{H}{T}, \qquad \frac{dy}{ds} = \frac{V}{T}, \qquad (V = \omega s, \quad H = \omega a),$$

$$T = \sqrt{H^2 + V^2} = \omega\sqrt{a^2 + s^2}.$$

The system of equations which describe the curve is

$$\frac{dx}{ds} = \frac{a}{\sqrt{a^2 + s^2}}, \qquad \frac{dy}{ds} = \frac{s}{\sqrt{a^2 + s^2}}. \tag{2}$$

Integration of these equations gives

$$x = a \sinh^{-1} \frac{s}{a}, \qquad y = \sqrt{a^2 + s^2}, \tag{3}$$

where both constants of integration are zero because of the way the coordinates were chosen. The parameter s can be eliminated (Problem 2) from equations (3) to give

$$y = a \cosh \frac{x}{a}. \tag{4}$$

Problems

1. Solve the system (2) and obtain (3). [*Hint:* Use the substitution $s = a \sinh \theta$ to integrate the first equation.]

2. Eliminate the parameter s from equations (3) to obtain (4). [*Hint:* Solve both equations for s^2.]

3. In Example 1, the arc length s of the curve from $A(0, a)$ to $B(x, y)$ is given by $s = \int_0^x \sqrt{1 + y'^2}\, dx$ $(y' = dy/dx)$. Show that $y' = V/H = s/a$ and that the curve can be described by $ay' = \int_0^x \sqrt{1 + y'^2}\, dx$. Differentiate this equation and solve the resulting second order equation.

4. A curve through $(0, 1)$ is such that the slope of the tangent line at any point (x, y), $x \geq 0$, is equal to twice the length of the arc from $(0, 1)$ to (x, y). Find the equation of the curve. Find the coordinates of the point on the curve whose distance from $(0, 1)$ along the curve is one unit.

$$\left[Hint: \quad \left(\frac{dx}{ds}\right)^2 + \left(\frac{dy}{ds}\right)^2 = 1 \quad \text{and} \quad \frac{dy/ds}{dx/ds} = \frac{dy}{dx} = 2s. \right]$$

5. A marble is propelled up in a vacuum at an angle α from the horizontal, with initial velocity v. Find the position at time t and show the path is parabolic.

6. A bomb of mass m is dropped from an airplane flying horizontally at speed v. The horizontal and vertical components of the force of air resistance are respectively $a(dx/dt)^2$ and $a(dy/dt)^2$, where a is constant. Take the origin as the drop point, with the positive y-axis pointing down, and show that the trajectory is described by the system

$$\frac{d^2x}{dt^2} = -\frac{a}{m}\left(\frac{dx}{dt}\right)^2, \qquad \left(x = 0, \ \frac{dx}{dt} = v \ \text{ when } t = 0\right),$$

$$\frac{d^2y}{dt^2} = g - \frac{a}{m}\left(\frac{dy}{dt}\right)^2, \qquad \left(y = \frac{dy}{dt} = 0 \ \text{ when } t = 0\right).$$

Find x and y at time t.

7. An object slightly heavier than water is released under water by a submarine moving horizontally at small velocity v (so there is nonturbulent flow through a somewhat viscous medium). The force of water resistance in this case has components $a(dx/dt)$ and $a(dy/dt)$. If the object has mass m, and weight in water W, find the parametric equations of the path.

ANSWERS

3. $y = a \cosh \left(\dfrac{x}{a} \right)$

4. $x = \frac{1}{2} \sinh^{-1} 2, \quad y = \dfrac{1}{2} + \dfrac{\sqrt{5}}{2}; \quad y = \frac{1}{2}(1 + \cosh 2x)$

5. $x = (v \cos \alpha)t, \quad y = (v \sin \alpha)t - \frac{1}{2}gt^2$

6. $x = \dfrac{m}{a} \ln \left[\dfrac{a}{m} vt + 1 \right], \quad y = \dfrac{m}{a} \ln \cosh \left(\sqrt{\dfrac{ag}{m}}\, t \right)$

7. $x = \dfrac{mv}{a} (1 - e^{-(a/m)t}), \quad y = \dfrac{Wm}{a^2} (e^{-(a/m)t} - 1) + \dfrac{W}{a} t$

7–3 A system equivalent to $M(x, y)\, dx + N(x, y)\, dy = 0$. We have
seen that the solutions of the system

$$\frac{dx}{dt} = F(t, x, y),$$

$$\frac{dy}{dt} = G(t, x, y) \tag{1}$$

can be interpreted as a family of plane curves given in parametric form.
We are familiar with the fact that the solutions of

$$M(x, y)\, dx + N(x, y)\, dy = 0 \tag{2}$$

form a family of curves. It is natural to ask whether one can always find
a parametric system (1) which has the same solution curves as any given
equation (2). If so, we would expect the system in some cases to be
easier to solve than (2), since many curves are more simply described
parametrically than by an equation in x and y. We can in fact show that
the equation (2) is equivalent to the system

$$\frac{dx}{N(x, y)} = \frac{dy}{-M(x, y)} = dt$$

which we agree is the same as

$$\frac{dx}{dt} = N(x, y),$$

$$\frac{dy}{dt} = -M(x, y). \tag{3}$$

Here "equivalent" means that every solution curve $y = h(x)$ of (2) has
a parametric representation $x = f(t)$, $y = g(t)$ which constitutes a solu-

tion of (3), and *vice versa*. Formally, this says that if $x = f(t)$, $y = g(t)$ is a solution of (3), and h is a function such that $g(t) \equiv h(f(t))$, then $y = h(x)$ is a solution of (2). Conversely, if $y = h(x)$ is a solution of (2), then there is a solution $x = f(t)$, $y = g(t)$ of (3) such that $g(t) \equiv h(f(t))$. These facts are proved in the following theorem.

THEOREM 1. *If f and g satisfy*

$$f'(t) \equiv N(f(t), g(t)),$$
$$g'(t) \equiv -M(f(t), g(t)), \tag{4}$$

and

$$g(t) \equiv h(f(t)),$$

then h satisfies

$$M(x, h(x)) + N(x, h(x)) \, h'(x) \equiv 0. \tag{5}$$

Conversely, if h satisfies (5), then there are functions f and g satisfying (4) such that $g(t) \equiv h(f(t))$.

Proof. The first part of the proof is left to the student (Problem 1). Let us show that every solution h of (2) has a parametric representation satisfying (4). Assume therefore that h satisfies the identity (5). Let f be a solution of the differential equation

$$f'(t) = N(f(t), h(f(t))). \tag{6}$$

Define g by the equation $g(t) = h(f(t))$. With this definition of g, (6) is the first of equations (4), and we need only verify the second. From the chain rule we have $g'(t) = h'(f(t))f'(t)$. Now rewrite the identity (5) by substituting $f(t)$ for x; by definition of g we can then write $g(t)$ for $h(x) = h(f(t))$, and $g'(t)/f'(t)$ for $h'(x) = h'(f(t))$. The result is

$$M(f(t), g(t)) + N(f(t), g(t)) \frac{g'(t)}{f'(t)} \equiv 0.$$

By rearranging and comparing with (6), we get

$$\frac{f'(t)}{N(f(t), g(t))} \equiv \frac{g'(t)}{-M(f(t), g(t))} \equiv 1.$$

Therefore g satisfies the second of equations (4). We have found a solution (f, g) of (3) such that $x = f(t)$, $y = g(t)$ is a parametric representation of the curve $y = h(x)$.

EXAMPLE 1. $(2x - y)\, dx + \sqrt{1 - x^2}\, dy = 0$. The equivalent system can be written

$$\frac{dx}{\sqrt{1 - x^2}} = -\frac{dy}{2x - y} = dt,$$

or

$$\frac{dx}{dt} = \sqrt{1 - x^2}, \qquad \frac{dy}{dt} = y - 2x. \tag{7}$$

The variables separate in the first equation, and we can integrate to get

$$\sin^{-1} x = t + c_1, \qquad x = \sin(t + c_1).$$

Substitution in the second equation gives

$$\frac{dy}{dt} - y = -2 \sin(t + c_1).$$

There is a particular solution of this linear equation of the form $y = A \cos(t + c_1) + B \sin(t + c_1)$, and we determine by substitution that $A = B = 1$. The solutions of the system (7) are therefore

$$y = c_2 e^t + \cos(t + c_1) + \sin(t + c_1),$$
$$x = \sin(t + c_1).$$

Using the relations

$$t = \sin^{-1} x - c_1,$$
$$\cos(t + c_1) = \sqrt{1 - \sin^2(t + c_1)} = \sqrt{1 - x^2},$$

we can eliminate t to obtain

$$y = c e^{\sin^{-1} x} + \sqrt{1 - x^2} + x, \tag{8}$$

where $c = c_2 e^{-c_1}$.

PROBLEMS

1. Prove that if f and g satisfy (4) and $g(t) \equiv h(f(t))$, then h satisfies (5).

2. The equation of Example 1 can be written $y' = (y - 2x)/\sqrt{1 - x^2}$. For what points (a, b) is there a solution y such that $y(a) = b$? Check, by substitution, that the functions (8) are solutions of the equation. Use the uniqueness theorem (Theorem 4 of Section 6–3) to show that the family (8) contains all solutions.

3. List the various techniques of Chapter 1 for solving a first order equation and see whether any of them apply to the equation of Example 1.

Solve Problems 4 through 6 by the methods of Chapters 1 and 2, and also by writing as a system. Reconcile the answers by eliminating the parameter from the solution of the system.

4. $xy \, dx + e^{-x} \, dy = 0$.

5. $(e^x + y) \, dx + e^{-x} \, dy = 0$. [*Hint:* To integrate $\int e^{e^x} e^{2x} \, dx$, let $u = e^x$, $dx = (1/u) \, du$.]

6. $[-(x^4 + y^2)/x^2] \, dx + (y/x) \, dy = 0$. [*Hint:* To solve directly, try the substitution $y = ux$.]

Write an equivalent system and solve.

7. $\left(2y + \dfrac{1}{y}\right) dx + (x + 2y^2) \, dy = 0$.

8. $(y + \sin^{-1}x) \, dx + \sqrt{1 - x^2} \, dy = 0$.

9. $(y - 2x) \, dx + (1 + y) \, dy = 0$.

10. $(2y - 2y^2) \, dx + \left(2x - \dfrac{1}{y} + 1\right) dy = 0$

   [*Hint:* Let $u = \dfrac{1}{y}$, $\dfrac{du}{dt} = -\dfrac{1}{y^2} \, (dy/dt)$.]

ANSWERS

2. Any point (a, b) with $|a| < 1$

4. $x = \ln(t + c_1)$, $\ln|y| = (t + c_1)[1 - \ln(t + c_1)] + c_2$;
   $\ln|y| = (1 - x)e^x + c$

5. $x = \ln(t + c_1)$, $y = c_2 e^{-t} - t - c_1 + 1$;
   $y = ce^{-e^x} - e^x + 1$ $(c = c_2 e^{c_1})$

6. $x = c_1 e^t + c_2 e^{-t}$, $y = x(c_1 e^t - c_2 e^{-t})$;
   $y^2 = x^2(x^2 + c)$, $(c = -4c_1 c_2)$

7. $2y^2 + 1 = c_1 e^{4t}$, $x = -\frac{1}{5}c_1 e^{-4t} - 1 + c_2 e^{-t}$

8. $x = \sin(t + c_1)$, $y = c_2 e^{-t} \pm t \mp 1 \pm c_1$

9. $x = c_1 e^t + c_2 e^{-2t} - \frac{1}{2}$, $y = c_1 e^t - 2c_2 e^{-2t} - 1$

10. $y = (c_1 e^{2t} + 1)^{-1}$, $x = (c_2 - c_1 t)e^{2t}$

7-4 Existence and uniqueness theorems.

In this section we prove an existence and uniqueness theorem for a pair of simultaneous first order equations in two functions. The theorem states that for appropriate F and G there is one and only one pair of functions (y, z), defined on some interval around a, which satisfy the equations

$$y'(x) = F(x, y(x), z(x)), \qquad y(a) = b,$$
$$z'(x) = G(x, y(x), z(x)), \qquad z(a) = c. \tag{1}$$

The method of proof is the Picard method of Chapter 6, with only the modifications required to treat two equations simultaneously.

We first rewrite the system (1) as an equivalent pair of integral equations:

$$y(x) = b + \int_a^x F(t, y(t), z(t)) \, dt,$$

$$z(x) = c + \int_a^x G(t, y(t), z(t)) \, dt. \tag{2}$$

Next we define inductively two sequences $\{y_n\}$ and $\{z_n\}$ by the equations

$$y_0(x) = b,$$
$$z_0(x) = c,$$
$$y_1(x) = b + \int_a^x F(t, b, c) \, dt,$$
$$z_1(x) = c + \int_a^x G(t, b, c) \, dt, \tag{3}$$
$$\vdots$$
$$y_{n+1}(x) = b + \int_a^x F(t, y_n(t), z_n(t)) \, dt,$$
$$z_{n+1}(x) = c + \int_a^x G(t, y_n(t), z_n(t)) \, dt.$$

The proof consists in showing that these sequences $\{y_n\}$ and $\{z_n\}$ converge and that the limit functions constitute a solution. If there are limits y and z of the respective sequences $\{y_n\}$ and $\{z_n\}$ such that

$$\lim_{n \to \infty} \int_a^x F(t, y_n(t), z_n(t)) \, dt = \int_a^x F(t, y(t), z(t)) \, dt,$$

$$\lim_{n \to \infty} \int_a^x G(t, y_n(t), z_n(t)) \, dt = \int_a^x G(t, y(t), z(t)) \, dt, \tag{4}$$

it is clear that these limits satisfy (2), because of the defining relation (3).

As in the proof of Picard's theorem for a single equation, there are three things which must be checked to make a proof along the lines indicated. We must show that

    I. The scheme (3) defines all y_n and z_n on some common interval I_0 around a.

    II. The sequences $\{y_n\}$ and $\{z_n\}$ converge on I_0 to some limits y and z.

    III. The integrals in (4) converge as indicated for all x in I_0.

We state next the hypotheses on F and G under which we can verify the conditions I, II, and III; these assumptions will be used throughout this section.

The functions F and G are assumed to be continuous on some cube S with center at (a, b, c):

$$S = \{(x, y, z) : |x - a| \leq h, |y - b| \leq h, |z - c| \leq h\}.$$

Since F and G are continuous on the closed bounded set S, they are bounded on S. Let M be a number such that $1 \leq M$, and $|F(x, y, z)| < M$, $|G(x, y, z)| < M$ for all (x, y, z) in S. Let $h_0 = h/M$, so that $h_0 \leq h$, and $Mh_0 = h$. The interval I_0 of conditions I, II, and III will be $[a - h_0, a + h_0]$. We assume that F and G are not only continuous, but also satisfy the following Lipschitz condition on S: for some number A and any two points (x, y, z) and (x, y_1, z_1) in S,

$$|F(x, y, z) - F(x, y_1, z_1)| \leq A\{|y - y_1| + |z - z_1|\},$$
$$|G(x, y, z) - G(x, y_1, z_1)| \leq A\{|y - y_1| - |z - z_1|\}. \tag{5}$$

Now we proceed with the several parts of the proof.

THEOREM 1 (Condition I). *For each n, the functions y_n and z_n given by (3) are defined on I_0, and for every x in I_0, the point $(x, y_n(x), z_n(x))$ is in S.*

Proof. (By induction.) The functions y_0 and z_0 are obviously defined on I_0, and the points $(x, y_0(x), z_0(x)) = (x, b, c)$ are in S if x is in I_0. Suppose that y_n and z_n are defined on I_0 for some n, and that $(x, y_n(x), z_n(x))$ is in S for all x in I_0. Since F and G are continuous on the points $(t, y_n(t), z_n(t))$, t in I_0, the integrals in (3) defining $y_{n+1}(x)$ and $z_{n+1}(x)$ exist for all x in I_0; that is, y_{n+1} and z_{n+1} are defined on I_0. Using the fact that $|F(x, y, z)| < M$ for (x, y, z) in S, we have, for every x in I_0,

$$|y_{n+1}(x) - b| = \left| \int_a^x F(t, y_n(t), z_n(t))\, dt \right|$$
$$\leq M|x - a| \leq Mh_0 = h.$$

Similarly,

$$|z_{n+1}(x) - c| \leq h$$

for all x in I_0, and the point

$$(x, y_{n+1}(x), z_{n+1}(x))$$

is in S.

THEOREM 2 (Condition II). *The sequences $\{y_n\}$ and $\{z_n\}$ converge uniformly on I_0.*

Proof. We will actually demonstrate the uniform convergence of the two series whose respective nth partial sums are y_n and z_n. That is, we write

$$y_n(x) = y_0(x) + [y_1(x) - y_0(x)] + \cdots + [y_n(x) - y_{n-1}(x)],$$

$$z_n(x) = z_0(x) + [z_1(x) - z_0(x)] + \cdots + [z_n(x) - z_{n-1}(x)] \tag{6}$$

and prove that the following series converge uniformly on I_0.

$$\sum_{n=1}^{\infty} [y_{n+1}(x) - y_n(x)];$$

$$\sum_{n=1}^{\infty} [z_{n+1}(x) - z_n(x)]. \tag{7}$$

The Lipschitz condition (5) is used to estimate the size of the terms of the series (7) as follows:

$$\begin{aligned}
|y_2(x) - y_1(x)| &= \left| \int_a^x [F(t, y_1(t), z_1(t)) - F(t, y_0(t), z_0(t))] \, dt \right| \\
&\leq \left| \int_a^x A \left\{ |y_1(t) - y_0(t)| + |z_1(t) - z_0(t)| \right\} dt \right| \\
&\leq \left| \int_a^x A \{ h + h \} \, dt \right| \\
&= 2Ah|x - a|.
\end{aligned}$$

The same argument on G shows that

$$|z_2(x) - z_1(x)| \leq 2Ah|x - a|.$$

Using the inequalities just proved, we estimate $|y_3(x) - y_2(x)|$ and $|z_3(x) - z_2(x)|$:

$$\begin{aligned}
|y_3(x) - y_2(x)| &= \left| \int_a^x [F(t, y_2(t), z_2(t)) - F(t, y_1(t), z_1(t))] \, dt \right| \\
&\leq \left| \int_a^x A \left\{ |y_2(t) - y_1(t)| + |z_2(t) - z_1(t)| \right\} dt \right| \\
&\leq \left| \int_a^x A \{ 4Ah|t - a| \} \, dt \right| \\
&= \frac{4A^2 h}{2!} |x - a|^2.
\end{aligned}$$

Again the same argument on G shows that

$$|z_3(x) - z_2(x)| \leq \frac{4A^2 h}{2!} |x - a|^2.$$

Continuing in this way, we see that for every n and every x in I_0, we have

$$|y_{n+1}(x) - y_n(x)| \leq \frac{2^n A^n h}{n!} |x - a|^n \leq h \frac{(2Ah)^n}{n!},$$

$$|z_{n+1}(x) - z_n(x)| \leq \frac{2^n A^n h}{n!} |x - a|^n \leq h \frac{(2Ah)^n}{n!}.$$

(8)

Since the series $\sum_{n=1}^{\infty} B^n/n!$ converges for every number B, the estimates (8) show that both series (7) converge uniformly on I_0. This is the same as saying that the sequences of partial sums $\{y_n\}$ and $\{z_n\}$ converge uniformly on I_0.

THEOREM 3 (Condition III). *If y and z are the limits of the sequences $\{y_n\}$ and $\{z_n\}$ defined by (3), then for all x in I_0*

$$\lim_{n \to \infty} \int_a^x F(t, y_n(t), z_n(t)) \, dt = \int_a^x F(t, y(t), z(t)) \, dt,$$

and

$$\lim_{n \to \infty} \int_a^x G(t, y_n(t), z_n(t)) \, dt = \int_a^x G(t, y(t), z(t)) \, dt.$$

Proof. We treat the first equation above, writing it in the form

$$\lim_{n \to \infty} \int_a^x [F(t, y_n(t), z_n(t)) - F(t, y(t), z(t))] \, dt = 0.$$

(9)

It is sufficient to show that the integrand in (9) can be made uniformly small on I_0 by taking n sufficiently large. From the Lipschitz condition (5), we get

$$|F(t, y_n(t), z_n(t)) - F(t, y(t), z(t))|$$
$$\leq A\{|y_n(t) - y(t)| + |z_n(t) - z(t)|\}.$$

Since the sequences $\{y_n\}$ and $\{z_n\}$ converge uniformly on I_0 to y and z, the right side above will be uniformly small on I_0 for all sufficiently large n. The second equality of the theorem follows from the same argument applied to G.

This completes the proof that there is a solution (y, z) of (1) on the interval I_0. Notice that we can specify a minimum length for I_0, in terms of the bound M for F and G on S, and that we can characterize the functions y and z as limits of constructable sequences. It remains to be shown that the pair of functions found above is the *only* solution of (1) on I_0.

In order to apply the Lipschitz condition in the uniqueness proof, we need the following lemma, which states that any solution curve of (1) must stay within S, for x in I_0.

LEMMA 1. *If (y_1, z_1) is any solution of* (1) *or* (2) *on I_0, then* $(x, y_1(x), z_1(x))$ *is in S for all x in I_0.*

Proof. If $(x, y_1(x), z_1(x))$ is not in S for some x in I_0—say some $x > a$ —then by the continuity of y_1 and z_1, there is a smallest number x_1 greater than a such that $(x_1, y_1(x_1), z_1(x_1))$ is on the boundary of S. That is, $|y_1(x_1) - b| = h$ or $|z_1(x_1) - c| = h$. To be specific, suppose that $|y_1(x_1) - b| = h$ and for all x between a and x_1, $|y_1(x) - b| < h$ and $|z_1(x) - c| < h$. By the Mean Value Theorem, there is a number x_0 between a and x_1 such that

$$h = |y_1(x_1) - b| = |y_1(x_1) - y_1(a)| = |y_1'(x_0)||x_1 - a|. \quad (10)$$

Since (y_1, z_1) is assumed to be a solution of the system (1), we have, in particular,

$$y_1'(x_0) = F(x_0, y_1(x_0), z_1(x_0)),$$

and hence $|y_1'(x_0)| < M$. Therefore, from (10), we get

$$h = |y_1'(x_0)||x_1 - a| < M|x_1 - a| \leq Mh_0 = h.$$

In other words, the contradiction $h < h$ follows from the assumption that $(x, y_1(x), z_1(x))$ lies outside S for some x in I_0.

THEOREM 4 (Uniqueness of the solution). *If (y, z) and (y_1, z_1) are any solutions on I_0 of* (1) *or* (2), *then $y = y_1$ and $z = z_1$ on I_0.*

Proof. The proof is essentially the same as the proof of uniqueness for a single equation (cf. Theorem 4, Section 6–3). We show that if $y(x_0) = y_1(x_0)$ and $z(x_0) = z_1(x_0)$ for some point x_0 in I_0, then $y = y_1$ and $z = z_1$ and the interval $[x_0 - 1/(2A), x_0 + 1/(2A)]$, where A is the constant of the Lipschitz condition. Since $y(a) = y_1(a) = b$, and $z(a) = z_1(a) = c$, we can start the argument at $x_0 = a$, and repeat it a finite number of times to show that $y = y_1$ and $z = z_1$ on all of I_0.

Assume that $y(x_0) = y_1(x_0)$, and $z(x_0) = z_1(x_0)$ for some x_0 in I_0. Let $h(x) = y(x) - y_1(x)$, and $k(x) = z(x) - z_1(x)$, so that $h(x_0) = k(x_0) = 0$. We must show that $h(x) = 0$ and $k(x) = 0$ for all x in $[x_0 - 1/(2A), x_0 + 1/(2A)]$. By Lemma 1 and the Lipschitz condition on F and G, we obtain

$$\begin{aligned}
|h'(x)| &= |y'(x) - y_1'(x)| \\
&= |F(x, y(x)) - F(x, y_1(x))| \\
&\leq A|y(x) - y_1(x)| \\
&= A|h(x)|, \quad\quad\quad\quad\quad\quad\quad\quad (11)
\end{aligned}$$

and, similarly,

$$|k'(x)| \leq A|k(x)|. \tag{12}$$

Now let $|h(x_1)|$ and $|k(x_2)|$ be the maximum values that $|h(x)|$ and $|k(x)|$ assume on the closed interval $[x_0 - 1/(2A), x_0 + 1/(2A)]$. By the Law of the Mean,

$$|h(x_1)| = |h(x_1) - h(x_0)| = |h'(\xi_1)||x_1 - x_0|,$$

and

$$|k(x_2)| = |k(x_2) - k(x_0)| = |k'(\xi_2)||x_2 - x_0|$$

for some ξ_1 between x_0 and x_1, and some ξ_2 between x_0 and x_2. Using (11) and (12) with the last equalities, and the fact that $|x_1 - x_0| < 1/(2A)$ and $|x_2 - x_0| < 1/(2A)$, we get

$$|h(x_1)| \leq A|h(x_1)||x_1 - x_0| < \tfrac{1}{2}|h(x_1)|,$$

and

$$|k(x_2)| \leq A|k(x_2)||x_2 - x_0| < \tfrac{1}{2}|k(x_2)|.$$

Hence $|h(x_1)| = 0$ and $|k(x_2)| = 0$, and since the maximum values of $|h(x)|$ and $|k(x)|$ are zero, h and k are identically zero on $[x_0 - 1/(2A), x_0 + 1/(2A)]$.

This completes the proof that the system (1), for suitable functions F and G, has a unique solution on some interval around a. Our results are restated, in slightly more convenient form, in the following theorem.

THEOREM 5 (Summary). *If* F, G, F_y, F_z, G_y, G_z *are continuous in some cube containing* (a, b, c), *then there is, on some interval around* a, *one and only one solution* (y, z) *of the system*

$$y' = F(x, y, z),$$
$$z' = G(x, y, z)$$

such that $y(a) = b$ *and* $z(a) = c$.

Proof. The continuity of the partial derivatives of F and G implies that F and G satisfy the Lipschitz condition of the earlier theorems (Problem 5).

The system (1) is called a *linear system* if the functions F and G are linear in y and z; i.e., a linear system is one of the form

$$y' = p_{11}(x)y + p_{12}(x)z + q_1(x),$$
$$z' = p_{21}(x)y + p_{22}(x)z + q_2(x). \tag{13}$$

Linear systems allow the same sort of systematic treatment as single linear equations and constitute the most important type of system. The existence theorem for single linear equations, which is the foundation for all of Chapters 3 and 4, will be proved on the basis of the existence theorem for linear systems.

If p_{11}, p_{12}, p_{21}, p_{22}, q_1, and q_2 are all continuous on some interval I, then the system (13) satisfies the hypotheses of Theorem 5 for any (a, b, c) with a in I. However, we can obtain a much better result for (13) than for the general system (1). Theorem 5 says that (1) has a solution on *some* interval around a; this interval may be small even if the functions F and G and their partial derivatives are continuous everywhere (see Problem 7). For the linear system (13), we can show that there is a solution (y, z) defined on all of any interval on which the coefficient functions are continuous.

THEOREM 6. *If p_{11}, p_{12}, q_1, p_{21}, p_{22}, and q_2 are continuous on $[\alpha, \beta]$, and $\alpha < a < \beta$, and b, c are any numbers, then there is one and only one solution (y, z) of (13) on $[\alpha, \beta]$ such that $y(a) = b$ and $z(a) = c$.*

Proof. The proof is as before, only now we can show that the sequences $\{y_n\}$ and $\{z_n\}$ converge on the whole interval $[\alpha, \beta]$. The integrals (3) defining y_n and z_n are clearly defined for all x in $[\alpha, \beta]$ (Condition I). The rest of the proof of Theorem 5 was based only on the fact that the points $(x, y_n(x), z_n(x))$ were in S, where the Lipschitz condition on F and G was known to hold. For the linear case, the Lipschitz condition holds on any set of points (x, y, z) such that x is in $[\alpha, \beta]$ [Problem 8(b)]. The convergence required in Conditions II and III now follows on all of $[\alpha, \beta]$, as in the proofs of Theorems 2 and 3. The student is asked to supply the details of this argument in Problem 8.

PROBLEMS

1. Show that the system of integral equations (2) is equivalent to the system (1).

2. Solve the following system by finding $y_n(x)$, $z_n(x)$ for all n, and verifying conditions I, II, and III directly.

$$y' = z, \quad z' = y;$$
$$y(0) = z(0) = 1.$$

3. Find y_2 and z_2 for the system $y' = z - 1$, $z' = y - x$; $y(1) = 0$, $z(1) = 2$.

4. Make a formal proof by induction of the inequalities (8).

5. Prove Theorem 5 from the preceding theorems; i.e., show that the continuity of F_y, F_z, G_y, G_z on S (and in particular the boundedness of these functions) implies that F and G satisfy (5).

6. (a) Show that the function $G(x, y, z) = \frac{1}{3}z^{2/3}$ does not satisfy the Lipschitz condition of (5) on any cube S containing $(0, 0, 0)$. [*Hint:* If the condition holds on S, then $|G(0, 0, z) - G(0, 0, 0)|/|z - 0|$ is bounded on S.]

   (b) Find two solutions of the system $y' = yz$, $z' = \frac{1}{3}z^{2/3}$ such that $y(0) = z(0) = 0$.

7. Show that no solution of the following system is defined on an interval longer than $\pi/10$: $y' = 100 + y^2$, $z' = y$.

8. Assume that the functions p_{11}, p_{12}, q_1, p_{21}, p_{22}, q_2 of the linear system (13) are continuous on $[\alpha, \beta]$.

   (a) Show that the sequences $\{y_n\}$ and $\{z_n\}$ given by (3) are defined on $[\alpha, \beta]$.

   (b) Show that (5) is satisfied on $\{(x, y, z) : \alpha \le x \le \beta\}$.

   (c) Show that $\{y_n\}$ and $\{z_n\}$ converge uniformly on $[\alpha, \beta]$.

   (d) Show that the limits y and z of the sequences $\{y_n\}$ and $\{z_n\}$ are solutions of (13) on $[\alpha, \beta]$.

   (e) Show that there is only one solution of (13) on $[\alpha, \beta]$.

9. Show that y is a solution of the second order equation $y'' = G(x, y, y')$ if and only if (y, y') is a solution of the system $y' = z$, $z' = G(x, y, z)$. State and prove an existence and uniqueness theorem for $y'' = G(x, y, y')$, using Theorem 5.

10. State and prove an existence and uniqueness theorem for the linear equation $y'' + p_1y' + p_0y = q$, using Theorem 6.

ANSWERS

2. $y_n(x) = z_n(x) = 1 + x + x^2/2! + \cdots + x^n/n!$ $y(x) = z(x) = e^x$

3. $y_2(x) = \frac{3}{2}(x - 1) - \frac{1}{6}(x^3 - 1)$; $z_2(x) = 3 - x$

6. (b) $y = 0, z = 0; y = 0, z = x^3$

7. All solutions must satisfy $y = 10 \tan(10x + c)$,
   $z = -\ln|\cos(10x + c)|$, for some c.

7–5 Existence theorem for nth order equations. In this section we will extend the results of Section 7–4 to first order systems in n functions, and examine the connection between such systems and nth order equations in one function. The systems we consider are those of the form

$$
\begin{aligned}
y_1' &= F_1(x, y_1, \ldots, y_n), \\
y_2' &= F_2(x, y_1, \ldots, y_n), \\
&\vdots \\
y_n' &= F_n(x, y_1, \ldots, y_n).
\end{aligned}
\tag{1}
$$

A solution of (1) is an n-tuple of functions, $(y_1, \ldots, y_n)$, such that on some interval the following identities hold:

$$
y_k'(x) \equiv F_k(x, y_1(x), \ldots, y_n(x)), \qquad k = 1, \ldots, n.
$$

If the functions $F_1, \ldots, F_n$ are sufficiently well behaved near a point

$(a, b_1, \ldots, b_n)$, then there is, on some interval around a, exactly one solution $(y_1, \ldots, y_n)$ of (1) which satisfies the initial conditions

$$y_1(a) = b_1, \ldots, y_n(a) = b_n. \tag{2}$$

Theorem 5 is a precise formulation of this statement for the case $n = 2$. The existence theorem was proved for the special case $n = 2$, rather than for the general system (1), so that the notation could be kept as simple as possible, and the statements interpreted in familiar geometric terms. Except for the welter of subscripts required, the proof of the existence and uniqueness theorem for (1) is not essentially different from that given in Section 7–4. The approximating sequences $\{y_{1m}\}, \ldots, \{y_{nm}\}$ are defined inductively as before by the formulas

$$y_{k0}(x) = b_k,$$

$$y_{k,m+1}(x) = b_k + \int_a^x F_k(t, y_{1m}(t), \ldots, y_{nm}(t))\, dt \quad (k = 1, 2, \ldots, n). \tag{3}$$

By the same methods used in Section 7–4 these sequences can be shown to converge on some interval around a to functions $y_1, \ldots, y_n$ which constitute a solution of (1) satisfying the initial conditions (2). We will not repeat the details of the proof; the statement of the existence theorem for (1) is as follows:

THEOREM 1. *If $F_1, \ldots, F_n$ and the partial derivatives $(\partial/\partial y_j)F_k$ $(j, k = 1, \ldots, n)$ are continuous on the set of points $(x, y_1, \ldots, y_n)$ such that $|x - a| \leq h$, $|y_1 - b_1| \leq h, \ldots, |y_n - b_n| \leq h$, then there is one and only one solution $(y_1, \ldots, y_n)$ of (1) on some interval around a such that $y_1(a) = b_1, \ldots, y_n(a) = b_n$.*

A *linear* first order system in n functions is a system of the form

$$\begin{aligned}
y_1' &= p_{11}(x)y_1 + \cdots + p_{1n}(x)y_n + q_1(x),\\
y_2' &= p_{21}(x)y_1 + \cdots + p_{2n}(x)y_n + q_2(x),\\
&\ \ \vdots\\
y_n' &= p_{n1}(x)y_1 + \cdots + p_{nn}(x)y_n + q_n(x).
\end{aligned} \tag{4}$$

The proof given in Section 7–4 for the existence of solutions to a linear system in two functions also extends without essential change to the general case. We therefore have the following theorem for the linear system (4).

THEOREM 2. *If the functions p_{ij}, q_i, for $i, j = 1, \ldots, n$, are continuous on $[\alpha, \beta]$, and $a, b_1, \ldots, b_n$ are any numbers, with $\alpha < a < \beta$, then* (4) *has a unique solution* $(y_1, \ldots, y_n)$ *on* $[\alpha, \beta]$ *such that* $y_1(a) = b_1, \ldots,$ $y_n(a) = b_n$.

Using Theorem 1, we can now give a very simple proof of the existence theorem for a single nth order equation in one function (cf. Problem 9, Section 7–4). To show that the equation

$$y^{(n)} = F(x, y, y', \ldots, y^{(n-1)}) \tag{5}$$

has solutions, we recast it as a first order system in the functions y, $y', \ldots, y^{(n-1)}$. To conform with the notation of Theorem 1, we define functions $y_2, \ldots, y_n$ by

$$y' = y_2, \qquad y'' = y_3, \ldots, y^{(n-1)} = y_n. \tag{6}$$

With the agreement (6), y is a solution of (5) if and only if $(y, y_2, \ldots, y_n)$ is a solution of the system

$$\begin{aligned} y' &= y_2, \\ y_2' &= y_3, \\ &\,\,\vdots \\ y_n' &= F(x, y, y_2, \ldots, y_n). \end{aligned} \tag{7}$$

Initial conditions for (5)

$$y(a) = b_0, \qquad y'(a) = b_1, \ldots, y^{(n-1)}(a) = b_{n-1} \tag{8}$$

correspond to the following initial conditions for (7).

$$y(a) = b_0, \qquad y_2(a) = b_1, \ldots, y_n(a) = b_{n-1}. \tag{9}$$

The first $n - 1$ functions on the right of (7)—the functions $F_1, \ldots, F_{n-1}$ in the notation of Theorem 1—satisfy the hypotheses of Theorem 1 for any point $(a, b_0, \ldots, b_{n-1})$. Therefore (7) has a solution satisfying the initial conditions (9) provided F satisfies the condition of Theorem 1. It follows that (5) has a solution satisfying the initial conditions (8) if F satisfies this condition, and we have proved the following theorem.

THEOREM 3. *If F and its partial derivatives F_y, $F_{y'}, \ldots, F_{y^{(n-1)}}$ are continuous on the set of points $(x, y, y', \ldots, y^{(n-1)})$ such that $|x - a| \leq h$, $|y - b_0| \leq h$, $|y' - b_1| \leq h, \ldots, |y^{(n-1)} - b_{n-1}| \leq h$, then there is, on some interval around a, a unique solution y of (5) satisfying the initial conditions (8).*

Remark. There is a certain amount of notational confusion in the statement of the above theorem. Although the symbols $x, y, y', y'', \ldots,$

$y^{(n-1)}$ are not customarily used as dummy coordinates for a point in $(n + 1)$-dimensional space, they are used in this way in the statement of the theorem to conform with the appearance of F in Eq. (5). Thus F is a function of $n + 1$ variables, and $F_{y'}$, for example, means the partial derivative of F with respect to the third variable. The symbols y', y'', etc., in (8) refer as usual to the derivatives of the solution y.

The same sort of argument used to prove Theorem 3 from Theorem 1 can be used to deduce Theorem 4 below from Theorem 2.

THEOREM 4. *If p_0, p_1, . . . , p_{n-1}, q are continuous on $[\alpha, \beta]$, and a, b_0, b_1, . . . , b_{n-1} are any numbers with $\alpha < a < \beta$, then the linear equation*

$$y^{(n)} + p_{n-1}(x)y^{(n-1)} + \cdots + p_1(x)y' + p_0(x)y = q(x)$$

has a unique solution y on $[\alpha, \beta]$ such that $y(a) = b_0$, $y'(a) = b_1$, . . . , $y^{(n-1)}(a) = b_{n-1}$.

Proof. Problem 4.

Systems of higher order equations can also be treated as first order systems by introducing the derivatives as new functions. For example, the system

$$y' = F(x, y, z, z'),$$

$$z'' = G(x, y, z, z') \tag{10}$$

is equivalent to the first order system

$$z' = w,$$
$$y' = F(x, y, z, w),$$
$$w' = G(x, y, z, w).$$

It follows that if F, G and their partial derivatives with respect to the last three variables are continuous in a region containing the point (a, b, c, d), then there is a unique solution (y, z) of (10) on some interval around a such that $y(a) = b$, $z(a) = c$, and $z'(a) = d$.

PROBLEMS

1. Find the approximating functions $y_{10}, y_{20}, y_{30}; \ldots ; y_{13}, y_{23}, y_{33}$ as given by (3) for the system

$$y_1' = y_2, \qquad y_1(0) = 1,$$
$$y_2' = y_3, \qquad y_2(0) = 0,$$
$$y_3' = -y_1, \qquad y_3(0) = -1.$$

2. Find the approximating functions y_{13}, y_{23}, y_{33}, as in Problem 1, for the system

$$y_1' = y_2, \qquad y_1(0) = 1,$$
$$y_2' = y_1, \qquad y_2(0) = 0,$$
$$y_3' = y_2 + y_3, \qquad y_3(0) = 2.$$

3. For what numbers b_1, b_2, b_3 is there a solution (y_1, y_2, y_3) of the following system such that $y_1(1) = b_1$, $y_2(1) = b_2$, $y_3(1) = b_3$?

$$y_1' = x\sqrt{y_2} + y_3/y_1,$$
$$y_2' = \sin^{-1} y_3 + \sqrt{y_1 - x},$$
$$y_3' = x \ln (y_1 - y_2).$$

4. Use Theorem 2 to prove Theorem 4.

5. Write a first order system equivalent to

$$y'' = xz' + y^2,$$
$$z'' = y' + xz.$$

What initial conditions can be prescribed for y and z and their derivatives?

6. Write a first order system equivalent to

$$y'' = z' + yy',$$
$$z''' = xz^2 + y'z''.$$

What initial conditions can be prescribed for y and z and their derivatives?

7. State and prove, using Theorem 1, an existence and uniqueness theorem for the system

$$y'' = F(x, y, y', z, z'),$$
$$z'' = G(x, y, y', z, z').$$

8. Solve the system of Problem 2.

9. Solve the following system by first eliminating z to obtain a third order equation in y.

$$y'' = 3y + z,$$
$$z' = -2y.$$

10. Solve the following system by first eliminating z.

$$y'' = z - y',$$
$$z' = 2z - 2y.$$

ANSWERS

1. $y_{13} = 1 - \frac{1}{2}x^2 - \frac{1}{6}x^3$, $y_{23} = -x - \frac{1}{2}x^2$, $y_{33} = -1 - x + \frac{1}{6}x^3$

2. $y_{13} = 1 + \frac{1}{2}x^2$, $y_{23} = x + \frac{1}{6}x^3$, $y_{33} = 2 + 2x + \frac{3}{2}x^2 + \frac{1}{2}x^3$

3. $1 < b_1$, $0 < b_2 < b_1$, $|b_3| < 1$

5. $y' = u$, $z' = v$, $u' = xv + y^2$, $v' = u + xz$; $y(a)$, $y'(a)$, $z(a)$, $z'(a)$ can be prescribed arbitrarily for any a.

6. $y' = u$, $z' = v$, $v' = w$, $u' = v + yu$, $w' = xz^2 + uw$; $y(a)$, $y'(a)$, $z(a)$, $z'(a)$, $z''(a)$ can be prescribed arbitrarily for any a.

7. If F, G, and their derivatives with respect to y, y', z, and z' are continuous in a region containing (a, b, c, d, e), then there is a unique solution (y, z) such that $y(a) = b$, $y'(a) = c$, $z(a) = d$, $z'(a) = e$.

8. $y_1 = \cosh x$, $y_2 = \sinh x$, $y_3 = e^x + 1$

9. $y = c_1 e^{-2x} + c_2 e^x + c_3 x e^x$, $z = c_1 e^{-2x} - 2c_2 e^x + c_3(-2x + 2)e^x$

10. $y = c_1 e^x + c_2 e^{\sqrt{2}\,x} + c_3 e^{-\sqrt{2}\,x}$,
$z = 2c_1 e^x + (2 + \sqrt{2})c_2 e^{\sqrt{2}\,x} + (2 - \sqrt{2})c_3 e^{-\sqrt{2}\,x}$

7–6 Polygonal approximations for systems.

In this section we will give a method for obtaining polygonal approximations to the two functions y and z which constitute the solution of the system

$$y' = F(x, y, z), \qquad y(a) = b,$$
$$z' = G(x, y, z), \qquad z(a) = c. \tag{1}$$

Since the sequences $\{y_n\}$ and $\{z_n\}$ of the existence proof converge to the solution functions, we already have one method of obtaining approximate solutions of (1). The polygonal approximations, however, are generally easier to calculate than the Picard approximations, even though the latter are more convenient to use in the proof itself.

With each partition $\{a = x_0, x_1, \ldots, x_n = d\}$ of an interval $[a, d]$ there is associated a polygonal approximation to each of the functions y and z satisfying (1). Over the interval $[a, x_1]$ the approximations are just the lines tangent to the solution curves at (a, b) and (a, c); i.e., the lines through the points (a, b) and (a, c) with respective slopes $F(a, b, c)$ and $G(a, b, c)$. Let (x_1, y_1) and (x_1, z_1) be the points at which the two tangent lines intersect the vertical line $x = x_1$. Compute $F(x_1, y_1, z_1)$ and $G(x_1, y_1, z_1)$, and take the lines through (x_1, y_1) and (x_1, z_1) with these respective slopes for the approximations over the interval $[x_1, x_2]$. Continuing in this way we obtain a sequence of points (a, b), (x_1, y_1), $\ldots$, (x_n, y_n), where the y_i are approximate values for $y(x_i)$, and a similar sequence for the z-curve. The polygonal curves joining these vertices are the approximations to y and z for the given subdivision. At each stage we have (letting $y_0 = b$, $z_0 = c$)

$$y_{k+1} = y_k + F(x_k, y_k, z_k)(x_{k+1} - x_k),$$
$$z_{k+1} = z_k + G(x_k, y_k, z_k)(x_{k+1} - x_k). \tag{2}$$

The method outlined above extends in the natural way to give polygonal approximations to the functions in the solution of a first order system in n functions. Since a second or higher order equation in one function can be considered as a first order system, we can also use this method to obtain approximations to the solutions of such higher order equations. Consider, for example, the second order equation

$$y'' = G(x, y, y'), \qquad y(a) = b, \qquad y'(a) = c. \qquad (3)$$

We rewrite this as the system

$$\begin{aligned} y' &= z, & y(a) &= b, \\ z' &= G(x, y, z), & z(a) &= c, \end{aligned}$$

and use (2), with $F(x, y, z) = z$, to compute simultaneous approximations to y and y'.

EXAMPLE 1. $\qquad \begin{aligned} y' &= z - 2, & y(0) &= 1, \\ z' &= y + 3x, & z(0) &= 0. \end{aligned}$

We compute the approximations to y and z over the interval $[0, 1]$, using the partition $\{0, 0.2, 0.4, 0.6, 0.8, 1.0\}$. The calculations, to three decimal places, are shown in Table 1. The solutions of this system are $y = e^x - 3x, z = e^x - 1$. The approximate values from Table 1, rounded off to two decimal places, are listed in Table 2 with the correct values for comparison. The graphs of the solutions and the approximations found are shown in Fig. 7–3.

EXAMPLE 2. $y'' = x + y, y(0) = 1, y'(0) = 0$. We treat this equation in the same way as the system $y' = z, y(0) = 1; z' = x + y, z(0) = 0$. Let us again compute approximations on $[0, 1]$, using the partition $\{0, 0.2, 0.4, 0.6, 0.8, 1.0\}$. The calculations are listed in Table 3. The correct solutions, $y = e^x - x, z = y' = e^x - 1$, and the approximations found above are shown in Fig. 7–4.

TABLE 1

x_k	0	0.2	0.4	0.6	0.8	1.0
y_k	1	0.6	0.24	−0.072	−0.326	−0.511
z_k	0	0.2	0.44	0.728	1.074	1.489
$z_k - 2$	−2	−1.8	−1.56	−1.272	−0.926	
$\Delta y_k = \frac{1}{5}(z_k - 2)$	−0.4	−0.36	−0.312	−0.254	−0.185	
$y_k + \Delta y_k = y_{k+1}$	0.6	0.24	−0.072	−0.326	−0.511	
$y_k + 3x$	1	1.2	1.44	1.728	2.074	
$\Delta z_k = \frac{1}{5}(y_k + 3x)$	0.2	0.24	0.288	0.346	0.415	
$z_k + \Delta z_k = z_{k+1}$	0.2	0.44	0.728	1.074	1.489	

TABLE 2

x	0	0.2	0.4	0.6	0.8	1.0
y_k	1	0.60	0.24	-0.07	-0.33	-0.51
$y = e^x - 3x$	1	0.62	0.29	0.02	-0.17	-0.28
z_k	0	0.20	0.44	0.73	1.07	1.49
$z = e^x - 1$	0	0.22	0.49	0.82	1.23	1.72

TABLE 3

x_k	0	0.2	0.4	0.6	0.8	1.0
y_k	1	1.0	1.04	1.128	1.274	1.489
z_k	0	0.2	0.44	0.728	1.074	1.489
$\Delta y_k = \frac{1}{5}z_k$	0	0.04	0.088	0.146	0.215	
$y_k + \Delta y_k = y_{k+1}$	1	1.04	1.128	1.274	1.489	
$x_k + y_k$	1	1.2	1.44	1.728	2.074	
$\Delta z_k = \frac{1}{5}(x_k + y_k)$	0.2	0.24	0.288	0.346	0.415	
$z_k + \Delta z_k = z_{k+1}$	0.2	0.44	0.728	1.074	1.489	

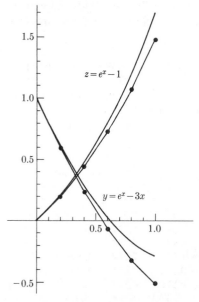

FIG. 7–3. Polygonal approximations for $y' = z - 2$, $z' = y + 3z$, $y(0) = 1$, $z(0) = 0$.

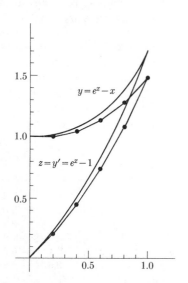

FIG. 7–4. Polygonal approximations for $y'' = x + y$, $y(0) = 1$, $y'(0) = 0$.

PROBLEMS

1. Compute approximations to the solutions of Example 1 using the partition $\{0, 0.1, 0.2, \ldots, 0.9, 1.0\}$. Compare your results with the correct values in Table 2.

2. Using the following partitions, find an approximation to the solution of the equation of Example 2.

(a) $\{0, 0.1, 0.2, \ldots, 0.9, 1.0\}$ of $[0, 1]$,

(b) $\{-0.5, -0.4, -0.3, -0.2, -0.1, 0\}$ of $[-\frac{1}{2}, 0]$.

3. For the system $y' = \frac{1}{2}z$, $z' = 2y$, $y(0) = \frac{1}{2}$, $z(0) = 1$, find approximations to y and z for the partition $\{0, 0.2, 0.4, 0.6, 0.8, 1.0\}$. Show that the solution is $y = \frac{1}{2}e^{-x}$, $z = -e^{-x}$, and plot the approximations and solution curves on the same graph.

4. Find approximations to the solution curves of

$$y' = 2(1 - z)/x^2, \qquad y(1) = 1,$$
$$z' = -3y, \qquad z(1) = 0,$$

(a) for the partition $\{0.2, 0.4, 0.6, 0.8, 1.0\}$,

(b) for the partition $\{0.2, 0.3, 0.4, 0.5, 0.6, 0.7, 0.8, 0.9, 1.0\}$.

Verify that the solution is $y = x^2$, $z = 1 - x^3$, and plot these curves with the approximations.

5. Plot approximations to the solution curves of the system of Problem 4 by using the partition $\{0.1, 0.2, 0.3, 0.4, 0.5\}$ of $[\frac{1}{10}, \frac{1}{2}]$.

(a) Start with the initial values $y(\frac{1}{2}) = 0.250$, $z(\frac{1}{2}) = 0.875$.

(b) Start with the initial values $y(\frac{1}{10}) = 0.010$, $z(\frac{1}{10}) = 0.999$.

6. Graph the polygonal approximations to y and y', where y is the solution of $y'' + y = 0$, $y(0) = 0$, $y'(0) = 1$, for the partition

$$\{0, 0.1, 0.2, \ldots, 1.5, 1.6, 1.7\}$$

of $[0, 1.7]$. What are the approximate values obtained for $y(\pi/2)$? for $y'(\pi/2)$?

ANSWERS

1. y_k: 1, 0.800, 0.610, 0.431, 0.264, 0.110, -0.029, -0.152, -0.257, -0.343, -0.407,

   z_k: 0, 0.100, 0.210, 0.331, 0.464, 0.610, 0.771, 0.948, 1.143, 1.357, 1.593

2. (a) y_k: 1, 1, 1.010, 1.031, 1.064, 1.110, 1.171, 1.248, 1.343, 1.457, 1.593,

   z_k: 0, 0.100, 0.210, 0.331, 0.464, 0.610, 0.771, 0.948, 1.143, 1.357, 1.593

   (b) y_k: 1, 1, 1.010, 1.029, 1.056, 1.090,

   z_k: 0, -0.100, -0.190, -0.271, -0.344, -0.410

3. y_k: 0.500, 0.600, 0.695, 0.784, 0.855, 0.919,

   z_k: 1.000, 0.950, 0.890, 0.820, 0.643, 0.557

4. (a) y_k: 1, 0.600, 0.350, 0.306, 0.731
      z_k: 0, 0.600, 0.960, 1.170, 1.354
   (b) y_k: 1, 0.800, 0.627, 0.483, 0.372, 0.301, 0.289, 0.383, 0.743,
      z_k: 0, 0.300, 0.540, 0.728, 0.873, 0.985, 1.075, 1.162, 1.277
5. (a) y_k: 0.250, 0.150, 0.087, 0.076, 0.181,
      z_k: 0.875, 0.950, 0.995, 1.021, 1.044
   (b) y_k: 0.010, 0.030, 0.050, 0.079, 0.114,
      z_k: 0.999, 0.996, 0.987, 0.972, 0.948
6. y_k: 0, 0.100, 0.200, 0.299, 0.396, 0.490, 0.580, 0.665, 0.744, 0.817, 0.882,
   0.939, 0.987, 1.026, 1.055, 1.074, 1.082, 1.079,
   z_k: 1, 1, 0.990, 0.970, 0.940, 0.900, 0.851, 0.793, 0.726, 0.652, 0.570, 0.482,
   0.388, 0.289, 0.186, 0.080, −0.027, −0.135
   At $\pi/2$, y-approximation is 1.080, y'-approximation is 0.004.

7-7 Linear systems. We return now to the linear system (4), Section 7-5, and indicate how the theory for such a system parallels that for a single linear equation (cf. Section 3-3). The system in question is

$$y_1' = p_{11}y_1 + \cdots + p_{1n}y_n + q_1,$$
$$\vdots \tag{1}$$
$$y_n' = p_{n1}y_1 + \cdots + p_{nn}y_n + q_n,$$

where the functions p_{ij} and q_i are assumed continuous on some interval $[\alpha, \beta]$. We know (Theorem 2, Section 7-5) that there is exactly one solution $(y_1, \ldots, y_n)$ of (1) on $[\alpha, \beta]$ such that $y_1(a) = b_1, \ldots, y_n(a) = b_n$ for any a in (α, β) and any numbers $b_1, \ldots, b_n$. If the functions q_i are all zero, then the system (1) is called *homogeneous*. The homogeneous system

$$y_1' = p_{11}y_1 + \cdots + p_{1n}y_n,$$
$$\vdots \tag{2}$$
$$y_n' = p_{n1}y_1 + \cdots + p_{nn}y_n$$

is called the *reduced* form of (1).

We define the *sum* of two solutions of (1) or (2), and a *constant multiple* of a solution as follows:

$$(y_{11}, \ldots, y_{1n}) + (y_{21}, \ldots, y_{2n}) = (y_{11} + y_{21}, \ldots, y_{1n} + y_{2n}),$$

$$c(y_1, \ldots, y_n) = (cy_1, \ldots, cy_n).$$

It then makes sense to talk of *linear combinations*

$$c(y_{11}, \ldots, y_{1n}) + \cdots + c_k(y_{k1}, \ldots, y_{kn})$$

of solutions. A set of solutions is *linearly independent* if no nontrivial linear combination is equal to the n-tuple $(0, \ldots, 0)$. The Wronskian of n solutions of (1) or (2) is the n-by-n determinant whose columns are the given solutions:

$$W(x) = \begin{vmatrix} y_{11}(x) & y_{21}(x) & \ldots & y_{n1}(x) \\ y_{12}(x) & y_{22}(x) & \ldots & y_{n2}(x) \\ \vdots & \vdots & & \vdots \\ y_{1n}(x) & y_{2n}(x) & \ldots & y_{nn}(x) \end{vmatrix}.$$

With the above definitions for sum, linear combination, Wronskian, etc., *all the theorems of Section 3–3 hold for the system* (1). To solve (1), therefore, it is necessary and sufficient to find one particular solution of (1), and all solutions of the reduced system (2). A linear combination of solutions of (2) is again a solution of (2). Any n solutions of (2) are linearly independent if and only if their Wronskian never vanishes. The set of all linear combinations of any n linearly independent solutions of (2) is the family of all solutions of (2). If $(y_{01}, \ldots, y_{0n})$ is a particular solution of (1), and $(y_{11}, \ldots, y_{1n}), \ldots, (y_{n1}, \ldots, y_{nn})$ are linearly independent solutions of (2), then every solution $(y_1, \ldots, y_n)$ of (1) can be expressed as follows.

$$\begin{aligned} y_1 &= y_{01} + c_1 y_{11} + \cdots + c_n y_{n1}, \\ y_2 &= y_{02} + c_1 y_{12} + \cdots + c_n y_{n2}, \\ &\vdots \\ y_n &= y_{0n} + c_1 y_{1n} + \cdots + c_n y_{nn}. \end{aligned} \tag{3}$$

Now we turn to the case in which the coefficients p_{ij} are constants, since this is the only case in which we have a systematic way of finding solutions to specific problems. We consider the homogeneous system (2), and illustrate the methods of finding solutions for the case $n = 2$:

$$\begin{aligned} y' &= p_{11}y + p_{12}z, \\ z' &= p_{21}y + p_{22}z. \end{aligned} \tag{4}$$

We must find two linearly independent solutions (y_1, z_1), (y_2, z_2), and by analogy with a single homogeneous linear equation, we try to find solutions of the form (Ae^{rx}, Be^{rx}). Substitution of $y = Ae^{rx}$, $z = Be^{rx}$ in (4) gives the equations

$$\begin{aligned} Are^{rx} &= p_{11}Ae^{rx} + p_{12}Be^{rx}, \\ Bre^{rx} &= p_{21}Ae^{rx} + p_{22}Be^{rx}. \end{aligned}$$

Hence (Ae^{rx}, Be^{rx}) is a solution of (4) if the numbers A, B, and r satisfy the following simultaneous equations:

$$(p_{11} - r)A + p_{12}B = 0,$$
$$p_{21}A + (p_{22} - r)B = 0. \tag{5}$$

This system of linear equations in A and B has a nontrivial solution if and only if the determinant equals zero; i.e., if the number r satisfies the equation:

$$\begin{vmatrix} p_{11} - r & p_{12} \\ p_{21} & p_{22} - r \end{vmatrix} = 0. \tag{6}$$

The quadratic equation (6) is called the *auxiliary equation* for the system (4). Suppose r_1 and r_2 are distinct roots of (6), and A_1, B_1 are numbers which satisfy (5) when $r = r_1$, and A_2, B_2 are numbers which satisfy (5) when $r = r_2$. Then $(A_1 e^{r_1 x}, B_1 e^{r_1 x})$ and $(A_2 e^{r_2 x}, B_2 e^{r_2 x})$ are solutions of (4), and every solution (y, z) of (4) can be written

$$y = c_1 A_1 e^{r_1 x} + c_2 A_2 e^{r_2 x},$$
$$z = c_1 B_1 e^{r_1 x} + c_2 B_2 e^{r_2 x}.$$

If the auxiliary equation has only one root r_0, then two solutions can be found by trying functions of the form $y = (A_1 + A_2 x)e^{r_0 x}$, $z = (B_1 + B_2 x)e^{r_0 x}$. If the roots of (6) are the complex numbers r_0 and $\bar{r}_0$, then there will be complex numbers A_0 and B_0 satisfying (5) when $r = r_0$. The pair $(A_0 e^{r_0 x}, B_0 e^{r_0 x})$ is a complex solution, and the real and imaginary parts are two real solutions.

The methods outlined above also extend to homogeneous systems in three or more functions (see Example 4).

EXAMPLE 1. $y' = y + 2z$, $z' = 2y + z$.

The auxiliary equation is

$$\begin{vmatrix} 1 - r & 2 \\ 2 & 1 - r \end{vmatrix} = r^2 - 2r - 3 = 0,$$

with roots $r = 3$, and $r = -1$. For $r = -1$ the equations (5) are both $2A + 2B = 0$, so we must have $A = -B$. We can take, for example, $A = 1$ and $B = -1$, and $(e^{-x}, -e^{-x})$ is one solution of the system. For $r = 3$ the equations (5) are both equivalent to $A = B$. Hence we have the second solution (e^{3x}, e^{3x}). Any solution (y, z) can therefore be expressed

$$y = c_1 e^{-x} + c_2 e^{3x},$$
$$z = -c_1 e^{-x} + c_2 e^{3x}.$$

EXAMPLE 2. $y' = 3y - 2z, \qquad z' = 2y - z.$

The auxiliary equation for this system is

$$\begin{vmatrix} 3 - r & -2 \\ 2 & -1 - r \end{vmatrix} = (r - 1)^2 = 0,$$

with the single root $r = 1$. Here we try solutions $y = (A_1 + A_2 x)e^x$, $z = (B_1 + B_2 x)e^x$. Substituting these functions in the system and dividing out e^x gives

$$A_1 + A_2 + A_2 x = 3A_1 - 2B_1 + (3A_2 - 2B_2)x,$$
$$B_1 + B_2 + B_2 x = 2A_1 - B_1 + (2A_2 - B_2)x.$$

For these equations to be identities, we must have

$$A_1 + A_2 = 3A_1 - 2B_1,$$
$$A_2 = 3A_2 - 2B_2,$$
$$B_1 + B_2 = 2A_1 - B_1,$$
$$B_2 = 2A_2 - B_2.$$

The second and fourth equations reduce to $A_2 = B_2$, and the first and third equations both become $2A_1 - A_2 - 2B_1 = 0$. We can take, for example, $A_1 = B_1 = 1$, $A_2 = B_2 = 0$, or $A_1 = 1$, $B_1 = 0$, $A_2 = B_2 = 2$. The solutions obtained for these values are (e^x, e^x) and $[(1 + 2x)e^x, 2xe^x]$. All solutions of the system can be expressed

$$y = c_1 e^x + c_2(1 + 2x)e^x = (c_1 + c_2 + 2c_2 x)e^x,$$
$$z = c_1 e^x + c_2 2x e^x = (c_1 + 2c_2 x)e^x.$$

EXAMPLE 3. $y' = y - z, \qquad z' = y + z.$

The auxiliary equation is $r^2 - 2r + 2 = 0$, with the complex roots $1 + i, 1 - i$. For $r = 1 + i$ the equations (5) both read

$$[1 - (1 + i)]A - B = 0,$$

or $A = iB$. Taking $B = 1$, $A = i$, we get the complex solution $(ie^{(1+i)x}, e^{(1+i)x})$, or

$$(ie^x \cos x - e^x \sin x, e^x \cos x + ie^x \sin x).$$

The real parts and imaginary parts must also be solutions, and we have the two real solutions

$$(-e^x \sin x, \ e^x \cos x), \qquad (e^x \cos x, \ e^x \sin x).$$

The family of all solutions is

$$y = -c_1 e^x \sin x + c_2 e^x \cos x,$$
$$z = c_1 e^x \cos x + c_2 e^x \sin x.$$

If the other root, $r = 1 - i$, of the auxiliary equation is used, one arrives at the same pair of solutions.

EXAMPLE 4. To illustrate how these methods extend to three equations in three functions, we consider the system

$$y' = y + w,$$
$$z' = 2y - z,$$
$$w' = 2w.$$

There are solutions of the form $(Ae^{rx}, Be^{rx}, Ce^{rx})$ for numbers r which satisfy

$$\begin{vmatrix} 1 - r & 0 & 1 \\ 2 & -1 - r & 0 \\ 0 & 0 & 2 - r \end{vmatrix} = -(1 - r)(1 + r)(2 - r) = 0.$$

We treat the roots $1, -1, 2$ in turn. The triple (Ae^x, Be^x, Ce^x) is a solution of the system if

$$0A + 0B + 1C = 0,$$
$$2A - 2B + 0C = 0,$$
$$0A + 0B + 1C = 0.$$

These equations are satisfied by $A = B = 1$, $C = 0$, and $(e^x, e^x, 0)$ is a solution. Similarly, if we substitute the functions $y = Ae^{-x}$, $z = Be^{-x}$, $w = Ce^{-x}$ in the system, we get the equations

$$2A + 0B + 1C = 0,$$
$$2A + 0B + 0C = 0,$$
$$0A + 0B + 3C = 0,$$

which are satisfied by $A = C = 0$, $B = 1$. Thus $(0, e^{-x}, 0)$ is a second solution of the system. For $r = 2$, the condition that $(Ae^{2x}, Be^{2x}, Ce^{2x})$ be a solution is

$$(-1)A + 0B + 1C = 0,$$
$$2A - 3B + 0C = 0,$$
$$0A + 0B + 0C = 0.$$

We must have $C = A$, and $B = \frac{2}{3}A$; taking $A = 3$ we get the third solution $(3e^{2x}, 2e^{2x}, 3e^{2x})$. Any solution (y, z, w) can be expressed as a linear combination of $(e^x, e^x, 0)$, $(0, e^{-x}, 0)$, and $(3e^{2x}, 2e^{2x}, 3e^{2x})$, and we can express the solutions

$$y = c_1e^x + 3c_3e^{2x}, \qquad z = c_1e^x + c_2e^{-x} + 2c_3e^{2x}, \qquad w = 3c_3e^{2x}.$$

PROBLEMS

Solve the following systems.

1. $y' = 2y + 3z,$
   $z' = 2y + z$

2. $y' = y,$
   $z' = -2z$

3. $y' = 2y + 5z,$
   $z' = y - 2z$

4. $y' = y,$
   $z' = 2y + z$

5. $y' = y - z$
   $z' = y + 3z$

6. $y' = y - z,$
   $z' = 2y - z$

7. $y' = y - 2z,$
   $z' = y + 3z$

8. $y' = y + 2w,$
   $z' = y + 2z + w,$
   $w' = 3y$

9. $y' = 2y - 2z - 4w,$
   $z' = 2y - 3z - 2w,$
   $w' = 4y - 2z - 6w.$ The auxiliary equation is $-(r + 2)^2(r + 3) = 0.$
Show that there are three linearly independent solutions of the form

$$(A_1e^{-2x}, B_1e^{-2x}, C_1e^{-2x}), \qquad (A_2e^{-2x}, B_2e^{-2x}, C_2e^{-2x}),$$
$$(A_3e^{-3x}, B_3e^{-3x}, C_3e^{-3x}).$$

ANSWERS

1. $c_1e^{-x} + 3c_2e^{4x}, \ z = -c_1e^{-x} + 2c_2e^{4x}$
2. $y = c_1e^x, \ z = c_2e^{-2x}$
3. $y = 5c_1e^{3x} - c_2e^{-3x}, \ z = c_1e^{3x} + c_2e^{-3x}$
4. $y = c_1e^x, \ z = c_1e^x + 2c_2xe^x$
5. $y = c_1e^{2x} + c_2xe^{2x}, \ z = -c_1e^{2x} - c_2(1 + x)e^{2x}$
6. $y = c_1 \cos x + c_2 \sin x, \ z = c_1 (\cos x + \sin x) + c_2 (\sin x - \cos x)$
7. $y = -2c_1e^{2x} \cos x - 2c_2e^{2x} \sin x$
   $z = c_1e^{2x} (\cos x - \sin x) + c_2e^{2x} (\cos x + \sin x)$
8. $y = c_2e^{3x} + 8c_3e^{-2x}, \ z = c_1e^{2x} + 2c_2e^{3x} + c_3e^{-2x}, \ w = c_2e^{3x} - 12c_3e^{-2x}$
9. $(0, -2e^{-2x}, e^{-2x}), \ (e^{-2x}, 2e^{-2x}, 0), \ (2e^{-3x}, e^{-3x}, 2e^{-3x});$
   $y = c_2e^{-2x} + 2c_3e^{-3x},$
   $z = -2c_1e^{-2x} + 2c_2e^{-2x} + c_3e^{-3x},$
   $w = c_1e^{-2x} + 2c_3e^{-3x}$

7–8 Operator methods. We have so far considered only systems of equations of the simple form in which each derivative is expressed explicitly in terms of the unknowns. It is frequently necessary to consider more general systems, in which the derivatives of several functions occur in each equation, and second or higher order derivatives appear. An example of such a system is

$$2y'' - 4y - z' = 4x,$$
$$2y' + 4z' - 3z = 0,$$

which can be conveniently written in operator notation as

$$(2D^2 - 4)y - Dz = 4x,$$
$$2Dy + (4D - 3)z = 0. \tag{1}$$

We consider systems, such as (1) above, which are linear in the unknowns and their derivatives and have constant coefficients. The general form of such a system is

$$L_{11}(D)y_1 + \cdots + L_{1n}(D)y_n = q_1,$$
$$L_{21}(D)y_1 + \cdots + L_{2n}(D)y_n = q_2,$$
$$\vdots \tag{2}$$
$$L_{n1}(D)y_1 + \cdots + L_{nn}(D)y_n = q_n,$$

where the $L_{ij}(D)$ are linear operators with constant coefficients, and the q_i are arbitrary continuous functions.

Since our existence theorem (Theorem 2, Section 7–5) is stated for linear systems in the simple form (4), Section 7–5, which we will call the *basic form*, we ask when the system (2) is equivalent to a system of this basic form. If we can solve (2) algebraically for the highest order derivatives of $y_1, \ldots, y_n$ which occur in the system, we can write an equivalent system in the basic form by introducing the lower order derivatives as new unknowns. To illustrate with the system (1), consider this as a linear *algebraic* system in D^2y and Dz:

$$2D^2y - Dz = 4y + 4x,$$
$$4Dz = -2Dy + 3z. \tag{3}$$

The determinant of this system is

$$\begin{vmatrix} 2 & -1 \\ 0 & 4 \end{vmatrix} = 8 \neq 0,$$

so we can solve for D^2y and Dz, and we get

$$D^2y = -\tfrac{1}{4}Dy + 2y + \tfrac{3}{8}z + 2x,$$
$$Dz = -\tfrac{1}{2}Dy + \tfrac{3}{4}z.$$

If we let $Dy = w$, whence $D^2y = Dw$, we obtain the following system in basic form which is equivalent to (1):

$$Dy = w,$$
$$Dw = -\tfrac{1}{4}w + 2y + \tfrac{3}{8}z + 2x, \tag{4}$$
$$Dz = -\tfrac{1}{2}w + \tfrac{3}{4}z.$$

It follows that (1) has solutions and that arbitrary initial conditions can be prescribed for y, Dy, and z. The general solution of (4) will contain linear combinations of three linearly independent solutions of the reduced system. Therefore the general solution of (1) or (3) will have y and z expressed as linear combinations involving three arbitrary constants.

If the system (2) cannot be solved algebraically for the highest order derivatives which occur, it is not equivalent to a system in basic form and may in fact be inconsistent. For example, the following system is inconsistent.

$$(D^2 + 2D)y + (2D - 3)z = e^x,$$
$$(D^2 + 2D)y + (2D - 3)z = 0.$$

We will treat only those systems which can be put in the basic form, the so-called *nondegenerate systems*.

Our methods of solving systems such as (1) frequently introduce extraneous functions. That is, we obtain formulas for y and z which are necessary conditions that (y, z) be a solution, but not sufficient. In practice, this means that the formulas contain more constants than they should, and it is convenient to have a check on how many arbitrary constants should appear. There will be as many arbitrary constants as there are functions in the equivalent basic system, which introduces a new function for each derivative occurring in the original system except those of highest order. For example, suppose that D^3y and D^2z are the highest order derivatives which appear in a system of two equations in y and z. The equivalent basic system involves the five functions y, $y_2 = Dy$, $y_3 = D^2y$, z, $z_2 = Dz$, and there must be five arbitrary constants in the solution. In general, *the number of constants in the solution is the sum of the highest orders of the derivatives which occur in the system.*

The preceding discussion indicates generally what sort of solutions the system (2) can be expected to have. Now we turn to the problem of finding the solutions. We proceed in much the same way as with algebraic systems. From the given system, we obtain new equations by differentiating and multiplying by constants, etc., so these equations can be combined to yield an equation in just one of the unknowns. This equation is solved and the solution substituted in the system. The procedure is illustrated in the examples below.

EXAMPLE 1.

$$y' = -y + z,$$
$$z' = -5y + 3z. \tag{5}$$

This system can be written in operator form

$$(D + 1)y - z = 0, \tag{6}$$
$$5y + (D - 3)z = 0. \tag{7}$$

Operating on (6) with $(D - 3)$, we obtain

$$(D - 3)(D + 1)y - (D - 3)z = 0. \tag{8}$$

Adding (8) and (7), and simplifying, we get

$$[(D - 3)(D + 1) + 5]y = 0, \tag{9}$$
$$(D^2 - 2D + 2)y = 0. \tag{10}$$

The solutions of (10) are

$$y = c_1 e^x \cos x + c_2 e^x \sin x.$$

From (6), $z = (D + 1)y$, or

$$z = c_1 e^x(2 \cos x - \sin x) + c_2 e^x(2 \sin x + \cos x).$$

Here the Eqs. (10) and (6) form a system equivalent (Problem 1) to the original system (6) and (7), and there is no problem of extraneous solutions. Note that the formulas for y and z contain two constants, as they should.

EXAMPLE 2.

$$(D - 1)y + (D + 1)z = e^{-x}, \tag{11}$$
$$D^2y + Dz = 2e^{-x}.$$

To eliminate z, operate on the first equation with D and on the second with $(D + 1)$ to obtain

$$D(D - 1)y + D(D + 1)z = -e^{-x},$$
$$(D + 1) D^2y + (D + 1) Dz = 0. \tag{12}$$

Subtracting the corresponding sides of equations (12), we get

$$[(D + 1)D^2 - D(D - 1)]y = e^{-x},$$
$$(D^3 + D)y = D(D^2 + 1)y = e^{-x}. \tag{13}$$

The solutions of (13) are

$$y = c_1 + c_2 \cos x + c_3 \sin x - \tfrac{1}{2}e^{-x}. \tag{14}$$

From the second of equations (11), we have

$$Dz = 2e^{-x} - D^2y = \tfrac{5}{2}e^{-x} + c_2 \cos x + c_3 \sin x,$$
$$z = -\tfrac{5}{2}e^{-x} + c_2 \sin x - c_3 \cos x + c_4. \tag{15}$$

Since D^2y and Dz are the highest order derivatives which appear, there should be $2 + 1 = 3$ arbitrary constants instead of four. The functions y and z of (14) and (15) satisfy the second equation of (11) for all values of c_1, c_2, c_3, and c_4. However, substitution in the first equation of (11) shows that (y, z) is a solution if and only if $c_1 = c_4$. Hence the solutions of (11) are given by

$$y = c_1 + c_2 \cos x + c_3 \sin x - \tfrac{1}{2}e^{-x},$$
$$z = c_1 + c_2 \sin x - c_3 \cos x - \tfrac{5}{2}e^{-x}.$$

EXAMPLE 3. We solve the system (1). Operating on the first equation with $(4D - 3)$, and on the second with D, we get

$$(4D - 3)(2D^2 - 4)y - (4D - 3) Dz = 16 - 12x,$$
$$2 D^2y + D(4D - 3)z = 0.$$

Adding these equations and simplifying, we obtain

$$2[(4D - 3)(D^2 - 2) + D^2]y = 16 - 12x,$$
$$(2D^3 - D^2 - 4D + 3)y = 4 - 3x,$$
$$(2D + 3)(D - 1)^2y = 4 - 3x.$$

The solutions of this equation are

$$y = -x + c_1e^x + c_2xe^x + c_3e^{-(3/2)x}.$$

From the first equation of (1), we have

$$\begin{aligned} Dz &= (2D^2 - 4)y - 4x \\ &= -2c_1e^x + c_2(-2x + 4)e^x + \tfrac{1}{2}c_3e^{-(3/2)x}. \end{aligned}$$

Hence z must satisfy

$$z = -2c_1e^x + c_2(-2x + 6)e^x - \tfrac{1}{3}c_3e^{-(3/2)x} + c_4.$$

Substituting y and z in the second of equations (1) gives $3c_4 + 2 = 0$, so $c_4 = -\tfrac{2}{3}$. Accordingly the solutions are given by

$$y = -x + c_1e^x + c_2xe^x + c_3e^{-(3/2)x},$$

$$z = -\tfrac{2}{3} - 2c_1e^x + c_2(-2x + 6)e^x - \tfrac{1}{3}c_3e^{-(3/2)x}.$$

Problems

1. Show that the system consisting of equations (10) and (6) is equivalent to the system of equations (6) and (7). [Since (10) was derived from (6) and (7), all that remains is to assume (10) and (6), and derive (7)].

2. Solve the system (5) of Example 1 by means of the auxiliary equation, as in Section 7–7.

3. Substitute the formulas (14) and (15) in the system (11) of Example 2 and show that $c_1 = c_4$ is necessary for (y, z) to be a solution.

4. Show that the following system is nondegenerate. How many constants must appear in the general solution?

$$\begin{aligned} D^2y + (D + 1)z + Dw &= e^x, \\ (2D + 1)y + 2z + D^2w &= 1, \\ (D^3 + D^2)y + (D + 2)z + (D - 3)w &= x. \end{aligned}$$

5. (See Example 3.) Show that $(y, z, w) = (-x, -\tfrac{2}{3}, -1)$ is a particular solution of the basic system (4) which is equivalent to (1). Show that

$$(e^x, -2e^x, e^x), \qquad (xe^x, (-2x + 6)e^x, (x + 1)e^x),$$

and

$$(e^{-(3/2)x}, -\tfrac{1}{3}e^{-(3/2)x}, -\tfrac{3}{2}e^{-(3/2)x})$$

are solutions of the reduced form of (4).

Solve the following systems.

6. $(D - 1)y + z = e^x,$
   $\quad -2y + (D + 1)z = 1$

7. $D^2y + (2D + 3)z = 3x,$
   $\quad Dy + Dz = x - 1$

8. $(D - 1)y - Dz = e^x,$ 9. $(D - 1)y + (D^2 - 1)z = 3e^{2x},$
   $2Dy - (D + 1)z = 2e^x$ $D^2y + (D + 1)z = 0$

10. Show that the following system is degenerate and that there is exactly one pair of functions which satisfies the system

$$(D^2 - 2)y + Dz = 2x,$$
$$3Dy + 3z = 1.$$

11. Show that the following system is inconsistent.

$$(D^2 - D)y + (D + 1)z + (D^3 - D^2)w = e^x,$$
$$(D + 1)y + (D - 1)z - (D^3 + D)w = 1,$$
$$(D^2 + 1)y + 2Dz - (D^2 + D)w = 2x.$$

Show that the system is consistent if the right member of the last equation is changed to $e^x + 1$. What can you say about the solutions?

ANSWERS

2. $(e^{(1+i)x}, (2 + i)e^{(1+i)x})$ is a complex solution.
4. six
6. $y = \frac{1}{2}c_1 (\cos x - \sin x) + \frac{1}{2}c_2 (\cos x + \sin x) + e^x - 1,$
   $z = c_1 \cos x + c_2 \sin x + e^x - 1$
7. $y = \frac{1}{2}x^2 - 2x - c_1e^{3x} - c_2e^{-x} + c_3,$
   $z = x - 1 + c_1e^{3x} + c_2e^{-x}$
8. $y = c_1 \cos x + c_2 \sin x,$
   $z = c_1 (\cos x - \sin x) + c_2 (\sin x + \cos x) - e^x$
9. $y = -e^{2x} + c_1e^{-x} + c_2e^x + c_3xe^x,$
   $z = \frac{4}{3}e^{2x} - c_1xe^{-x} - \frac{1}{2}c_2e^x - c_3(\frac{1}{2}x + \frac{3}{4})e^x + c_4e^{-x}$
10. $y = -x,\ z = \frac{4}{3}$
11. If w is any function, there is a three-parameter family of solutions for y and z in terms of w.

7–9 Laplace transform methods. The Laplace transform provides an effective way of treating systems of linear equations with constant coefficients, as well as single equations in one unknown. By taking transforms, we convert a system of linear equations in y and z into an algebraic system of linear equations in $\hat{y}$ and $\hat{z}$. If the original system is consistent, this algebraic system can be solved for $\hat{y}$ and $\hat{z}$, and then y and z can be determined by the methods of Chapter 5. As for a single equation, the Laplace transform method gives directly the particular solution which corresponds to a given set of initial conditions.

In this section we will illustrate how the transform method applies to some of the examples of the preceding sections.

EXAMPLE 1. (cf. Example 1, Section 7–7.) We consider the system

$$y' = y + 2z,$$
$$z' = 2y + z, \tag{1}$$

with the initial conditions

$$y(0) = 2, \qquad z(0) = 0. \tag{2}$$

Taking transforms of both sides of equations (1), we get

$$s\hat{y} - 2 = \hat{y} + 2\hat{z},$$
$$s\hat{z} - 0 = 2\hat{y} + \hat{z},$$

or

$$(s - 1)\hat{y} - 2\hat{z} = 2,$$
$$-2\hat{y} + (s - 1)\hat{z} = 0. \tag{3}$$

The determinant of the system (3) is

$$\begin{vmatrix} s - 1 & -2 \\ -2 & s - 1 \end{vmatrix} = s^2 - 2s - 3 = (s - 3)(s + 1).$$

Hence $\hat{y}$ and $\hat{z}$ are given by

$$\hat{y} = \frac{1}{(s - 3)(s + 1)} \begin{vmatrix} 2 & -2 \\ 0 & s - 1 \end{vmatrix} = \frac{2s - 2}{(s - 3)(s + 1)},$$

$$\hat{z} = \frac{1}{(s - 3)(s + 1)} \begin{vmatrix} s - 1 & 2 \\ -2 & 0 \end{vmatrix} = \frac{4}{(s - 3)(s + 1)}.$$

From the last equation above and #7 of Table 2 (Section 5–4), we get

$$z = \frac{4}{3 - (-1)} [e^{3x} - e^{-x}] = e^{3x} - e^{-x}.$$

Using #8 and #7 of Table 2 in Section 5–4, and the formula above for $\hat{y}$, we get

$$y = \tfrac{2}{4}[3e^{3x} + e^{-x}] - \tfrac{2}{4}[e^{3x} - e^{-x}]$$
$$= e^{3x} + e^{-x}.$$

If, instead of the initial conditions (2), we take arbitrary initial conditions $y(0) = a$, $z(0) = b$, then the method above gives the general solution of the system (1) in the form (Problem 1)

$$y = \tfrac{1}{2}(a + b)e^{3x} + \tfrac{1}{2}(a - b)e^{-x},$$

$$z = \tfrac{1}{2}(a + b)e^{3x} - \tfrac{1}{2}(a - b)e^{-x}.$$

(4)

EXAMPLE 2. (cf. Example 4, Section 7–7.)

$$y' = y + w, \qquad y(0) = 4,$$
$$z' = 2y - z, \qquad z(0) = 2,$$
$$w' = 2w, \qquad w(0) = 3.$$

(5)

The transformed equations for the system (5) are

$$s\hat{y} - 4 = \hat{y} + \hat{w},$$
$$s\hat{z} - 2 = 2\hat{y} - \hat{z},$$
$$s\hat{w} - 3 = 2\hat{w},$$

which can be written

$$(s - 1)\hat{y} + 0\hat{z} - 1\hat{w} = 4,$$
$$-2\hat{y} + (s + 1)\hat{z} + 0\hat{w} = 2,$$
$$0\hat{y} + 0\hat{z} + (s - 2)\hat{w} = 3.$$

(6)

From the last of equations (6), we get

$$\hat{w} = \frac{3}{s - 2}.$$

Substitution of this result in the first equation gives

$$\hat{y} = \frac{4}{s - 1} + \frac{3}{(s - 2)(s - 1)} = \frac{1}{s - 1} + \frac{3}{s - 2}.$$

From the second of equations (6), we find that

$$\hat{z} = \frac{2}{s + 1} + \frac{2}{(s - 1)(s + 1)} + \frac{6}{(s - 2)(s + 1)}.$$

Decomposing the terms above into partial fractions gives

$$\hat{z} = \frac{1}{s - 1} - \frac{1}{s + 1} + \frac{2}{s - 2}.$$

With the formulas above for $\hat{w}$, $\hat{y}$, and $\hat{z}$, we can identify the solution as

$$y = e^x + 3e^{2x},$$
$$z = e^x - e^{-x} + 2e^{2x},$$
$$w = 3e^{2x}.$$

The next example shows how the transform method applies to the higher order systems treated in Section 7–8. Notice how the operator manipulations of Example 2, Section 7–8, parallel the algebraic operations in the following example.

EXAMPLE 3. (cf. Example 2, Section 7–8.)

$$y' - y + z' + z = e^{-x}, \qquad y(0) = 2, \qquad y'(0) = \tfrac{1}{2}$$
$$y'' + z' = 2e^{-x}, \qquad z(0) = -1 \tag{7}$$

The equations in $\hat{y}$ and $\hat{z}$ corresponding to (7) are

$$s\hat{y} - 2 - \hat{y} + s\hat{z} + 1 + \hat{z} = \frac{1}{s+1},$$
$$s^2\hat{y} - 2s - \tfrac{1}{2} + s\hat{z} + 1 = \frac{2}{s+1},$$

which we write as

$$(s-1)\hat{y} + (s+1)\hat{z} = \frac{1}{s+1} + 1 = \frac{s+2}{s+1},$$
$$s^2\hat{y} + s\hat{z} = \frac{2}{s+1} + 2s - \frac{1}{2} = \frac{1}{2}\frac{4s^2 + 3s + 3}{s+1}. \tag{8}$$

To solve for $\hat{y}$ and $\hat{z}$, multiply the first of equations (8) by s, and the second by $s + 1$; this gives

$$s(s-1)\hat{y} + s(s+1)\hat{z} = \frac{s^2 + 2s}{s+1},$$
$$s^2(s+1)\hat{y} + s(s+1)\hat{z} = \frac{1}{2}\frac{4s^3 + 7s^2 + 6s + 3}{s+1}. \tag{9}$$

Subtracting the corresponding sides of (9), we get

$$(s^3 + s)\hat{y} = s(s^2 + 1)\hat{y} = \frac{1}{2}\frac{4s^3 + 5s^2 + 2s + 3}{s+1},$$

or

$$\hat{y} = \frac{1}{2} \frac{4s^3 + 5s^2 + 2s + 3}{s(s+1)(s^2+1)} = \frac{\frac{3}{2}}{s} - \frac{\frac{1}{2}}{s+1} + \frac{s}{s^2+1}. \tag{10}$$

Hence

$$y = \tfrac{3}{2} - \tfrac{1}{2}e^{-x} + \cos x,$$

which is Formula (14), Section 7–8, with $c_1 = \frac{3}{2}$, $c_2 = 1$, and $c_3 = 0$. From the second of equations (8), and (10), we obtain

$$\hat{z} = -s\hat{y} + 2 + \frac{2}{s(s+1)} - \frac{\frac{1}{2}}{s}$$

$$= -\frac{1}{2} \frac{4s^3 + 3s^2 + 2s + 3}{(s+1)(s^2+1)} + 2 + \frac{2}{s(s+1)} - \frac{\frac{1}{2}}{s}$$

$$= -\frac{1}{2} \frac{s^2 - 2s - 1}{(s+1)(s^2+1)} + \frac{2}{s(s+1)} - \frac{\frac{1}{2}}{s}.$$

Breaking the terms above into partial fractions, we get

$$\hat{z} = \frac{-\frac{1}{2}}{s+1} + \frac{1}{s^2+1} + \frac{2}{s} - \frac{2}{s+1} - \frac{\frac{1}{2}}{s}$$

$$= \frac{-\frac{5}{2}}{s+1} + \frac{\frac{3}{2}}{s} + \frac{1}{s^2+1}.$$

Hence

$$z = -\tfrac{5}{2}e^{-x} + \tfrac{3}{2} + \sin x,$$

which is Eq. (15), Section 7–8, again with $c_2 = 1$, and $c_3 = 0$.

PROBLEMS

1. Use transforms, with the initial conditions $y(0) = a$, $z(0) = b$, to obtain the general solution (4) of the system (1).

2. Use transforms to solve the systems cited, for the given initial conditions:

  (a) System of Example 2, Section 7–7; $y(0) = 0$, $z(0) = -1$
  (b) System of Example 3, Section 7–7; $y(0) = 1$, $z(0) = 0$
  (c) System of Example 1, Section 7–8; $y(0) = 1$, $z(0) = 2$
  (d) Problem 6, Section 7–8; $y(0) = 0$, $z(0) = 0$
  (e) Problem 7, Section 7–8; $y(0) = 1$, $z(0) = -1$
  (f) Problem 8, Section 7–8; $y(0) = 0$, $z(0) = 0$
  (g) Problem 9, Section 7–8; $y(0) = -1$, $z(0) = \frac{4}{3}$

3. THEOREM: *For any transform $\hat{f}$, $\lim_{s\to\infty} \hat{f}(s) = 0$.*

Use this fact to show that the system of Problem 10, Section 7–8, has only one solution. [*Hint:* Let $y(0) = a$, $y'(0) = b$, $z(0) = c$ be any initial conditions; show that $\hat{y} = -1/s^2 + \frac{1}{6}[1 - 3b - 3c]$, and hence that $y = -x$, etc.]

ANSWERS

2. (a) $y = 2xe^x$, $x = -1 + 2xe^x$
   (b) $y = e^x \cos x$, $z = e^x \sin x$
   (c) $y = e^x \cos x$, $z = e^x(2 \cos x - \sin x)$
   (d) $y = e^x - 1$, $z = e^x - 1$
   (e) $y = \frac{1}{2}x^2 - 2x + 1$, $z = x - 1$
   (f) $y = \sin x$, $z = \sin x + \cos x - e^x$
   (g) $y = -e^{2x}$, $z = \frac{4}{3}e^{2x}$

INDEX

A CATALOG OF SELECTED
DOVER BOOKS
IN SCIENCE AND MATHEMATICS

Astronomy

BURNHAM'S CELESTIAL HANDBOOK, Robert Burnham, Jr. Thorough guide to the stars beyond our solar system. Exhaustive treatment. Alphabetical by constellation: Andromeda to Cetus in Vol. 1; Chamaeleon to Orion in Vol. 2; and Pavo to Vulpecula in Vol. 3. Hundreds of illustrations. Index in Vol. 3. 2,000pp. 6⅛ x 9¼.
23567-X, 23568-8, 23673-0 Three-vol. set

THE EXTRATERRESTRIAL LIFE DEBATE, 1750–1900, Michael J. Crowe. First detailed, scholarly study in English of the many ideas that developed from 1750 to 1900 regarding the existence of intelligent extraterrestrial life. Examines ideas of Kant, Herschel, Voltaire, Percival Lowell, many other scientists and thinkers. 16 illustrations. 704pp. 5⅜ x 8½. 40675-X

A HISTORY OF ASTRONOMY, A. Pannekoek. Well-balanced, carefully reasoned study covers such topics as Ptolemaic theory, work of Copernicus, Kepler, Newton, Eddington's work on stars, much more. Illustrated. References. 521pp. 5⅜ x 8½.
65994-1

AMATEUR ASTRONOMER'S HANDBOOK, J. B. Sidgwick. Timeless, comprehensive coverage of telescopes, mirrors, lenses, mountings, telescope drives, micrometers, spectroscopes, more. 189 illustrations. 576pp. 5⅜ x 8¼. (Available in U.S. only.)
24034-7

STARS AND RELATIVITY, Ya. B. Zel'dovich and I. D. Novikov. Vol. 1 of *Relativistic Astrophysics* by famed Russian scientists. General relativity, properties of matter under astrophysical conditions, stars, and stellar systems. Deep physical insights, clear presentation. 1971 edition. References. 544pp. 5⅜ x 8¼. 69424-0

Chemistry

CHEMICAL MAGIC, Leonard A. Ford. Second Edition, Revised by E. Winston Grundmeier. Over 100 unusual stunts demonstrating cold fire, dust explosions, much more. Text explains scientific principles and stresses safety precautions. 128pp. 5⅜ x 8½. 67628-5

THE DEVELOPMENT OF MODERN CHEMISTRY, Aaron J. Ihde. Authoritative history of chemistry from ancient Greek theory to 20th-century innovation. Covers major chemists and their discoveries. 209 illustrations. 14 tables. Bibliographies. Indices. Appendices. 851pp. 5⅜ x 8½. 64235-6

CATALYSIS IN CHEMISTRY AND ENZYMOLOGY, William P. Jencks. Exceptionally clear coverage of mechanisms for catalysis, forces in aqueous solution, carbonyl- and acyl-group reactions, practical kinetics, more. 864pp. 5⅜ x 8½.
65460-5

Physics

OPTICAL RESONANCE AND TWO-LEVEL ATOMS, L. Allen and J. H. Eberly. Clear, comprehensive introduction to basic principles behind all quantum optical resonance phenomena. 53 illustrations. Preface. Index. 256pp. 5⅜ x 8½.   65533-4

ULTRASONIC ABSORPTION: An Introduction to the Theory of Sound Absorption and Dispersion in Gases, Liquids and Solids, A. B. Bhatia. Standard reference in the field provides a clear, systematically organized introductory review of fundamental concepts for advanced graduate students, research workers. Numerous diagrams. Bibliography. 440pp. 5⅜ x 8½.   64917-2

QUANTUM THEORY, David Bohm. This advanced undergraduate-level text presents the quantum theory in terms of qualitative and imaginative concepts, followed by specific applications worked out in mathematical detail. Preface. Index. 655pp. 5⅜ x 8½.   65969-0

ATOMIC PHYSICS (8th edition), Max Born. Nobel laureate's lucid treatment of kinetic theory of gases, elementary particles, nuclear atom, wave-corpuscles, atomic structure and spectral lines, much more. Over 40 appendices, bibliography. 495pp. 5⅜ x 8½.   65984-4

AN INTRODUCTION TO HAMILTONIAN OPTICS, H. A. Buchdahl. Detailed account of the Hamiltonian treatment of aberration theory in geometrical optics. Many classes of optical systems defined in terms of the symmetries they possess. Problems with detailed solutions. 1970 edition. xv + 360pp. 5⅜ x 8½.   67597-1

THIRTY YEARS THAT SHOOK PHYSICS: The Story of Quantum Theory, George Gamow. Lucid, accessible introduction to influential theory of energy and matter. Careful explanations of Dirac's anti-particles, Bohr's model of the atom, much more. 12 plates. Numerous drawings. 240pp. 5⅜ x 8½.   24895-X

ELECTRONIC STRUCTURE AND THE PROPERTIES OF SOLIDS: The Physics of the Chemical Bond, Walter A. Harrison. Innovative text offers basic understanding of the electronic structure of covalent and ionic solids, simple metals, transition metals and their compounds. Problems. 1980 edition. 582pp. 6⅛ x 9¼.   66021-4

HYDRODYNAMIC AND HYDROMAGNETIC STABILITY, S. Chandrasekhar. Lucid examination of the Rayleigh-Benard problem; clear coverage of the theory of instabilities causing convection. 704pp. 5⅜ x 8¼.   64071-X

INVESTIGATIONS ON THE THEORY OF THE BROWNIAN MOVEMENT, Albert Einstein. Five papers (1905–8) investigating dynamics of Brownian motion and evolving elementary theory. Notes by R. Fürth. 122pp. 5⅜ x 8½.   60304-0

THE PHYSICS OF WAVES, William C. Elmore and Mark A. Heald. Unique overview of classical wave theory. Acoustics, optics, electromagnetic radiation, more. Ideal as classroom text or for self-study. Problems. 477pp. 5⅜ x 8½.   64926-1

CATALOG OF DOVER BOOKS

METHODS OF THERMODYNAMICS, Howard Reiss. Outstanding text focuses on physical technique of thermodynamics, typical problem areas of understanding, and significance and use of thermodynamic potential. 1965 edition. 238pp. 5⅜ x 8½. 69445-3

TENSOR ANALYSIS FOR PHYSICISTS, J. A. Schouten. Concise exposition of the mathematical basis of tensor analysis, integrated with well-chosen physical examples of the theory. Exercises. Index. Bibliography. 289pp. 5⅜ x 8½. 65582-2

RELATIVITY IN ILLUSTRATIONS, Jacob T. Schwartz. Clear nontechnical treatment makes relativity more accessible than ever before. Over 60 drawings illustrate concepts more clearly than text alone. Only high school geometry needed. Bibliography. 128pp. 6⅛ x 9¼. 25965-X

THE ELECTROMAGNETIC FIELD, Albert Shadowitz. Comprehensive undergraduate text covers basics of electric and magnetic fields, builds up to electromagnetic theory. Also related topics, including relativity. Over 900 problems. 768pp. 5⅝ x 8¼. 65660-8

GREAT EXPERIMENTS IN PHYSICS: Firsthand Accounts from Galileo to Einstein, edited by Morris H. Shamos. 25 crucial discoveries: Newton's laws of motion, Chadwick's study of the neutron, Hertz on electromagnetic waves, more. Original accounts clearly annotated. 370pp. 5⅜ x 8½. 25346-5

RELATIVITY, THERMODYNAMICS AND COSMOLOGY, Richard C. Tolman. Landmark study extends thermodynamics to special, general relativity; also applications of relativistic mechanics, thermodynamics to cosmological models. 501pp. 5⅜ x 8½. 65383-8

LIGHT SCATTERING BY SMALL PARTICLES, H. C. van de Hulst. Comprehensive treatment including full range of useful approximation methods for researchers in chemistry, meteorology and astronomy. 44 illustrations. 470pp. 5⅜ x 8½. 64228-3

STATISTICAL PHYSICS, Gregory H. Wannier. Classic text combines thermodynamics, statistical mechanics and kinetic theory in one unified presentation of thermal physics. Problems with solutions. Bibliography. 532pp. 5⅜ x 8½. 65401-X

Paperbound unless otherwise indicated. Available at your book dealer, online at **www.doverpublications.com**, or by writing to Dept. GI, Dover Publications, Inc., 31 East 2nd Street, Mineola, NY 11501. For current price information or for free catalogues (please indicate field of interest), write to Dover Publications or log on to **www.doverpublications.com** and see every Dover book in print. Dover publishes more than 500 books each year on science, elementary and advanced mathematics, biology, music, art, literary history, social sciences, and other areas.